D0912600

THE GOLDEN THREAD OF TIME

A Quest for the Truth
and Hidden Knowledge of the Ancients

by
CRICHTON E M MILLER

Photographs by Maureen Palmer, Joyce McMillan & Crichton Miller
Illustrations by Crichton E M Miller using Apple Mac and Microsoft software
Technical Drawings by Chris Tillbrook

Pendulum Publishing

Pendulum Publishing

First published in Great Britain 2001 by

Pendulum Publishing
Rugby, Warwickshire, United Kingdom.

A CIP catalogue record for this title is available from the British Library.

ISBN 0-9541639-0-7

Designed & Typeset by Cre8 Communications Ltd,
Rugby, Warwickshire, United Kingdom.

Printed and bound in Great Britain by
Butler & Tanner Limited, Frome, Somerset.

CONTENTS

CHAPTER 1: THE FOUNDATION   1

CHAPTER 2: THE GOLDEN THREAD   21

CHAPTER 3: THE THIRD EYE   41

CHAPTER 7: GODS BROUGHT TO EARTH   171

CHAPTER 8: THE KNIGHTS TEMPLAR   233

CHAPTER 9: OVER THERE   267

THE GOLDEN THREAD OF TIME

"Chaos rules the world of Man and Nature
in the moment of now
Order will return for a while in the cycle of the times,
but the World will never die,
The Vine of Man and Woman may wither and pass away.

New Birth will come and Ma-at return.
New eyes may greet the rising of the Sun
In distant future, another kind may taste the living air of ancient dead
and rise in life from the dust of ages,
To wonder once again at the stars
and keep the Time to sail the seas on wind and tide.

They who are to walk the Earth
will spin beneath the ruthless eye of Starry Serpent
and watch the hoary Twelve sail by with Orion
at the helm
And will they see the Sun give life by day?
As did we,
and live in balance with the world
For a while.

What clothes will you wear, my love?
And how shall I recognise thee?
Will we be as One as once before?
When we meet again in this ship of life.
in distant times,
On the Golden Thread of millions of years."

© 2001 Crichton E M Miller

ACKNOWLEDGEMENTS

This book is dedicated to the force I personally know as The Great Spirit of Creation that you may perceive as God and who gave me all I see, hear, touch, smell, feel and think in this changing universe.
This book is also dedicated to Joyce my partner, who is my mortal guardian angel.

I give thanks to all my children who carry this knowledge into the future. Louise the Fierce, Annelies the Gentle, who enthusiastically accompanies me on field trips, Sebastian the Lion, my treasured son. Debbie the Scientist, Fiona the Sailor, and little Louise the Actress, who told me off once for not teaching her how to build an Ark for protection from the rising sea levels.

My respect and thank you to the "House of Ram", Aries, otherwise known as Amen, who left this treasure locked in stone over four thousand years ago as a celebration of time and a manifest multifunctional instrument incorporating the apex of their ancient science and spirituality. My heartfelt gratitude is that they built it strong, so that it might last and return to us the lost and forgotten knowledge of cyclical and calendrical mathematics with which our ancestors calculated future events.

My thanks go out to all those gifted, courageous writers and researchers for their discoveries, support and encouragement.

Maureen Palmer for her works on Simulacra and the contribution of an appendix within this publication.

Chris Ogylvie Herald for his support, friendship and encouragement over the years.

Ani Bealaura for her friendship and major contribution to the world through her hard work and amazing web site "The Gate of the Moon."

Greg Taylor for publishing my work on the Daily Grail website.

David Ritchie for understanding the mathematics of the ancients and adding a brilliant mathematical appendix for this book

Jim Bowles for writing the introduction to this work and understanding the nature of enlightenment.

Gary David for his discovery of another Orion correlation in America.

Simon Cox for his additional research on the Dead Sea Scrolls and Christopher Columbus.

My thanks go out to Robert Bauval for the kind words of encouragement that he gave me on seeing my discovery for the first time.

Chris Tillbrook and Jenny of Tillbrook and Co. Patent Attorneys, Warwick, England. For assisting me in achieving a patent on the greatest instrument of the ancients.

My thanks go out to the entire well wishing ordinary people and scientists alike; who have visited my website over the last year.

To the editors of The Coventry Evening Telegraph, Practical Boatowner and Hera magazine, for having the courage to print my controversial discovery.

A special thanks to my designers, Hamilton Ngan and Dawn Blaver at Cre8 Communications Ltd, Rugby, Warwickshire.

CHAPTER I

THE
FOUNDATION

ABOUT THE AUTHOR

My name is Crichton and I am a sovereign soul. By sovereign, I mean that I am a free thinker and not one who blindly follows seeing nothing for myself. I was given the name of Crichton before I was born (on the 25th of August 1949 in Glasgow), as were the entire first born sons in our ancient Scottish family. Crichton means Christ Town and is a Christian name. I have no religious background and I am a man just like any other. My body is formed from the material elements in the same way as yours are. I am subject to death and disease as you are, and I have suffered the trials and enjoyed the pleasures of life on this planet as you have done.

My window of life and space in this place and time, of the vine of life, is designated in the scheme of things as yours is.

I do not think that my credentials of study, or centre of learning, or my methods of earning a living are important, for these "*Labels*" have no bearing or relevance to the knowledge that I pass on to you. I am a qualified navigator and sailor and I am grateful for these skills that have helped me see through ancient eyes. I follow no creed but that of Truth and Wisdom, nor am I a member of any secret society, past or present. What I will pass on to you is another way of seeing the world of the Ancients and the world of Now through my eyes, experience and discovery. I will show you a world of ancient knowledge and understanding that has been left in wood and stone, in word and song over the centuries, by those who thought a different way.

I do not expect to change the world or save the many. I am not a religious zealot and this is not a book designed to prove either religion or history to be correct. Rather it is a work by an ordinary person trying to share that which he has discovered through reason and common sense. That reason and common sense will show religion and history, as we know it, to be incorrect. What I am about to reveal to you is a gift that was given to me by the ancients one winter night in 1997 after a lifetime of researching the truth about our history.

This book is the result of four years work and I hope you like it.

THE GIFT

The ancient knowledge that I will share in this book is a gift. A gift that will open the doors to further discoveries in others, as it sparks the imagination of the reader. It may help break the bonds of dogma and reveal the true knowledge that will help us understand our ancestors, the way they thought and bring us closer to a modern understanding of Nature, our World and the Cosmic cycles that affect our lives. This knowledge is a gift for our children and the future generations to come and for the dispossessed indigenous peoples of the world, who with great fortitude, have tried faithfully to keep the knowledge alive with their ceremonies, rituals and songs. It is a secret that has been carried down the centuries by Christians, Moslems, Knights Templar and Masons. It is a gift that will help dispel the icon as a path to God, especially as many religions forbid this practise for the blindness it brings.

This is a gift to help us see the invisible.

This is not another theory based on hard to prove ideas, its physical proof is an ancient measuring instrument that was used by our ancient ancestors before history was written. It is not a complicated tool, but is so simple as to be the mathematical embodiment of pure beauty and truth. In the right hands and with the right knowledge it is immensely powerful and changed the world in our past. It changed the world so greatly that the repercussions are still felt today after thousands of years. This instrument that I have discovered has become one of the most revered icons in our religious institutions and is known by all peoples around the world as a holy symbol. They are right, it is to be treasured and respected but not worshipped, for it is an object of great knowledge. It was an instrument that was developed from a wisdom that had immense power in its time. Its capability was such that the works it performed in the ancient past lead many modern groups of people to believe that aliens must have assisted in the achievements of the ancients. You will know that what I say and share is truth, for you already know what this instrument is and you will remember this ancient knowledge when you see it. You will see that the aliens are ourselves because we have forgotten what we once knew.

I have faith that my contribution to revealing these depths of ancient wisdom will inspire those who follow me to reach even greater levels of understanding and wisdom. I hope that this work may reveal a reason for the healing of difference and division within the Brother and Sisterhood of Mankind and Nature itself. The discoveries within will expose knowledge held by our ancient ancestors that, although different from our own and less technologically advanced, was fundamentally deeper in wisdom and understanding.

There are many people in the world who are dissatisfied with the teachings of conventional religion and feel that it does not answer the problems of a modern society. They are right, for the larger institutions have lost their way and are out of touch with the people. Many people also feel that the answers given by science do not satisfy a longing for something more than materialism in the human psyche. My work seeks a goal and points a finger from the dust of ages, to a future time where science and spirituality may become as one again as it was in the distant past. It is my wish that our unborn children will be in harmony with each other on this beautiful emerald planet. The planet called Earth, which is our home and the home of all the millions of creatures that dwell within its bounds.

I will show beyond reasonable doubt that ancient man was capable of a form of understanding of such depth and subtlety that we only glimpse it today, occasionally and with untrained eyes. Furthermore that this ancient knowledge could be as old as 30,000 years or more and may reveal information about our changing, dynamic world that could affect our descendants and those of us who are still living in the next few decades. You will realise, perhaps for the first time, the falsehood of the way in which we live and that mankind has not progressed despite the increasing pace of change and development of technology. You may well feel that in fact we are degenerating and that our system is in danger of collapse. I will supply as many references as I can at the end of this book, so as to assist in your further research, but as the reader must understand, original thought and discovery has few references. Where are the earlier references to the theory of relativity? This ancient instrument works, so it is not a theory. It is safe in my hands, I will not let it fall nor be used by those who seek the false glitter of games and power. It is real Gold of the kind that can not be lost or spent, but that which lifts the spirit and knowledge as one. I have discovered the Rose Stone, the Staff of Angles, the Time Machine of the Ancients, how they used it and some of their knowledge and the way they lived, flourished and died. What I have discovered, I have made, patented and protected. I have patented it so that it may not fall in the wrong hands and be used as another "Icon" type of symbol like the swastika to create division, hate and horror amongst the peoples of the world.

Of the many thousands of people, lay and academic alike, who have visited my website[i] and read my articles in various journals, I have received nothing but thoughtful and constructive messages of encouragement from complete strangers and friends alike. This is a book for the ordinary person and I have tried to make it as easy to understand as possible, supporting it with references to other books and web sites. There are academic references with this book as well for those who wish to go further.

My friend and fellow author, Jim Bowles[ii], will introduce my work with his special type of insight and views into the nature of how these kinds of discoveries occur to people like us. Jim has written many works of his own and possibly, his most memorable work is his remarkable book The Gods, Gemini and the Great Pyramid. Jim Bowles is a dedicated follower of Charles H. Hapgood[iii], who came up with the hypothesis of Earth crust displacement and had the support of Albert Einstein. Earth crust displacement has been put forward as an answer to the cause of the flood and the end of the last ice age. Jim has a background of engineering skills, which he gained as a rocket engineer for NASA; he has made some amazing discoveries, theories and insights into how the world may really work through repetitive cyclical changes.

FOREWORD

History doesn't write in the first person

By James Bowles[1]

My essay title, *History doesn't write in the first person*, is not intended to be confusing, rather it is a reminder that history does not write itself. It is written by those with the skills, the experience, and the interest; and by those with something to give and those with something to hide. We only have to recall the often expressed cliché, "History is recorded by the victor," to understand the hidden! But there's an altruistic side to hiding history as well that affects us deeply; it concerns a gift *(regalio)* and the giver, which we'll discuss in a moment.

First, let me explain why I chose to use "*the first person*" in the title.

Writing in the first person is my method of choice, because by way of example I can say *first hand*, "I have known Crichton Miller for a number of years now and have come to respect his integrity. One of the first contacts I had with him came about shortly after he'd discovered the significance of the cross and he wrote me expressing the hope that he had the mettle required of him in possessing, through discovery, an ancient wisdom." I wrote back saying, "Don't give it a thought Crichton, because there's a purity in the ancient wisdom that hides it from the self-indulgent. In a sense it's the Ancient's way of choosing who will carry on the tradition. And it appears that you've been selected."

I don't recall that this ancient methodology of passing on the tradition, which we modern people associate with serendipity, has ever been formally recognised. But I think it's worthy of a name, if for no other reason than it provides a means to put to memory those who have contributed *(through discovery)* to our knowledge of the lost ancient wisdoms.[2] By lost, I mean that there is nothing tangible, nothing conspicuous, about the existence of this lost knowledge; nothing tells us it's there. Rather, in discovery, it chooses us!

I don't believe we understand even the rudiments of this selection process, although we can certainly see its consequence. The sports field and the *Arts*, for instance (where a willingness to give of themselves is so conspicuous), is dominated by those with *innate abilities*. We lack a term for this too, and simply think of these people as *gifted*. The serendipitous, or chance discovery of a lost knowledge is likewise, and in no less way, a form of giftedness. Crichton specifically speaks of his discovery of the cross "as a flash of inspiration, a *gift* and a personal joy beyond belief in its revelation."

A seeming uniqueness of a serendipitous discovery, which may be part of the selection process, is an apparent responsibility. Crichton rarely speaks of his work without acknowledging this, saying, "I know that there is a purpose behind all of this, but I (also) know I am a servant." He speaks of the discovery as his having been given a *rare gift*, or present. In Italian, (because I like its fluidity) it would be said that he'd received *Uno caro regalio[3]* a rare gift! But we're speaking here too, as one having received the *regalio* and turning, as though the agent of transfer, and re-presenting it "Here I give you this *Uno caro regalio*."

Uno caro regalio is a phrase that we're less likely to use; so much as we'll remember and recognise for its associations. We take for granted, for instance, many extraordinary instances of discovery that might now be thought of in this way. Contributions made by names like Newton, Faraday and Maxwell who pioneered the field of physics when there was no physics to pioneer. They invented it, and now we as recipients, make use of it!

I'd also like to include Charles Hapgood in this gifted category. Charles Hapgood's name is currently shrouded in controversy. But I can assure you; you don't know what controversy is until you study the life and times of Faraday. Faraday's pioneering work on electromagnetism brought into existence the groundwork for the electrification of the entire world, although to this very day his name is all but ignored. Paradoxically, his contribution in the realm of electricity is so basic that nothing would work without it.

Hapgood's work, despite the errant controversy, was no less brilliant because he brought into existence a new field of geophysics, one that defined a world where temporary chaos initiated (though devastating to all life forms) a replenishing of the Earth's resources through crustal shift and cascading waters.

Returning now to the original theme for a moment, *History doesn't write in the first person*, means that you cannot walk up to the Great Pyramid and expect to find a plaque that says, "*You will find all the formulae, drawings and methods of construction on the thirtieth tier, under the thirty-first casing stone from the left.*" You'll no more find such a cache than you'll hear the winds and the waves carrying the sounds of the conversations of the past. It isn't there.

It means, further, that ferreting the ancient truths from the voiceless megaliths and metaphorical literature cannot be accomplished from walking the site, or reading the scripts. Rather it is a laborious task requiring all that seek to ignore the chants from established interests; that you discard the very fabric of modern thought, and place yourself close within their culture, understanding their means, and recognising their skills and admiring their goals. But, so as not to stumble on semantics here, let me define my term, *ancient truths*.

Ancient truths are the archetypal blueprints of the ancient culture, and are expressed in what they knew, what they believed, how they thought, how they lived and how they accomplished their tasks.

By definition, undiscovered ancient truths are outside of our understanding, there being no indication of their existence or vestigial form. Having said that, the point is immediately made partially mute by many researchers who have learned to employ the *archetypal blueprint* as a predictive tool. This term, archetypal blueprint, needs some clarification. Archetypal blueprint is not a frequently used term and is not well understood, so perhaps the best way from my point of view as an observer of the methodology rather than a studied user, is to summarise an article written by Dave Talbot entitled "The Thunderbolt In Myth And Symbol."[4]

Background

It'd be hard to count the times, while in discussions concerning the Ancients, that I'd heard the dissenting argument, (one thing or another) may be incorporated (in this construction or that), *but there's no proof they understood its principle?*" Or how many times have we encountered the argument "these monuments may be aligned with the stars of (such and such a constellation), but *there's no proof that this was their intent?*"

This form of argument has, if not by intent, than at least by effect, the consequence of perpetrating ignorance. Discussions like these tend to switch quickly from the subject of the skill, or the intent, to one of *"show me where they "said" they employed this or that."* Well unfortunately history is not written in the first person. There are no such statements of satisfaction. Where after all can you go, even today, and find anything so specific?

Metaphoric Styles

That challenge will always have to be dealt with I suppose, but it's not without response, particularly if we adopt the view, as many others and I have, that the Ancients (perhaps in anticipation of this human tendency) preserved their knowledge in metaphor. Metaphor was employed in the unique complexity of hieroglyphic text, and in the location and design dimensioning of the classical megalithic structures that we find throughout the world. Mathematical constants and principles of physics, astronomical alignments and holy precepts are all encoded in both styles of record, in a metaphoric style so complex that it hides its intent and thereby protects itself from purge.

Archetypal Blueprints using lightning as a comparitive example!

On the subject of my earlier suggestion, that it's *less than sufficient to walk the sites, or read the scripts*, Dave Talbott writes:[5]

"It is the recurring themes, the archetypes, that rescue us from scepticism, enabling us to distinguish the substratum of human memory from the carnival of fragmentation and elaboration over time. An archetype is an irreducible first form — it cannot be reduced to a more elementary statement; and (as far as can be determined from historical investigation), it has no precedent."

"Archetypes as a whole are the keys to our understanding (and rescue from the trials) of ancient myth-making[6] imagination."

"Our comparative investigation has identified hundreds of recurring themes of myth, including numerous global images of lightning. The disconnection of these images from the observed behaviour of lightning is an impressive anomaly. The usual tendency will be to look for what is intelligible under the tests of common experience. If, however, "uncommon experience" is the basis of the global imagery, then this very habit must be confronted as a prime obstruction to discovery."

"Let us begin with the most common ancient symbols of the divine thunderbolt. All of the unusual motifs listed below find wide distribution in the ancient world." (Author's note: For simplicity, I'm presenting only six of Dave Talbott's 21 original examples.)

Motif #1: Lightning takes the form of a frightful weapon. Often depicted as a sword, arrow, mace, club, spear, axe, or hammer.

Motif #2: Lightning appears as a great bird or "thunderbird" with heaven-spanning wings.

Motif #3: Lightning is a generative, masculine pillar. It impregnates goddesses.

Motif #4: Lightning is a "chain of arrows" launched skyward and appears as a ladder or backbone of the sky, whose steps were ascended by an ancestral hero.

Motif #5: Lightning "blossoms" as a flower, the celebrated plant of life.

Motif #6: Lightning is fire and brimstone (sulphur). The lightning of the gods.

Rules of Investigation

"A productive investigation of the archetypes require three overriding principles:

1) The investigation must focus exclusively on common mythical, symbolic or ritual themes: including all points of agreement between far-flung cultures.

2) Each verifiable theme must be traced to its earliest instances.

3) All common cultural expressions of the themes must be considered as evidence. Pictures illuminate ancient storytelling. Ritual celebrations give context to the pictures. Myths add crucial background to the rites. Certain extraordinary facts can now be stated concerning the archetypes, and these facts challenge all prior explanations or theories of myth."

From this new vantage point, it is now possible for the serious student to follow the progression of the symbolic language from first form, or archetype, through later elaboration."

History Doesn't Write in the First Person

Perhaps now we've gained the perspective to realise that my original statement is wrong. So is the assumption that the *"show me where they "said" that,"* argument is without response.

Because, now, in light of our recognition of the uniqueness and indisputable commonality of *innate ability*, of the similar commonality of *serendipitous discovery*, and the common opportunity for meaningful participation and contribution through the predictive power of the *archetype*; we can re-title this essay, *History DOES Write in the First Person* to those who have, are given, or work to acquire, *Uno caro regalio*, the gift to see what is concealed.

© 2001 James Bowles http://members.tripod.com/jamesbowles/

Endnotes:

1. James Bowles is the author of the book, The Gods, Gemini, and the Great Pyramid, articles, "30,000 Years Ago from yesterday," "Hapgood Revisited," "Wings of Thoth," and "Around the world and to the moon and back with Imsety, Hapy, Duamutef, and Kebhsenuf." Mr. Bowles also has a collection of narrative poetry in the five parts; "The Age of the Sphinx© " and "The Earth is My Ark©."

2. In reference to lost ancient wisdoms, it's often been expressed that the "science we see at the dawn of recorded history was not science at its dawning, but represents the remnants of the science of some great and as yet untracked civilisation." (Reference: Richard W. Noone, The Hammer and the Pendulum, NEXUS Magazine, Jan 1999.)

3. Connotations like colloquialisms are often dependent on local usage, so allow me to define what I mean by, Uno caro regalio. My usage means, "A gift of such unique value that I feel an obligation to offer it in return to others."

4. Dave Talbott, "The Thunderbolt In Myth And Symbol," "THOTH, Vol. V, No 4. March 15, 2001. Mikamar Publishing.

5. Dave Talbott's article, "The Thunderbolt in Myth and Symbol,"utilises the Thunderbolt *motif in its ancient representations of lightning as a demonstrable vehicle for his discussion of the archetype as a recurring irreducible first form.*

6. The reader should make a clear distinction here between myth making, *which stems from human interpretations, and the* irreducible first form *of the archetype. They stand apart as strongly as opposite poles, or we might observe that there is no myth in the archetypal form.*

THE CYCLES OF NATURE AND TIME

I do not have to prove to you that cycles are part of life and that these cycles of time affect us on a daily basis. The seasons change throughout the year as the Earth spins around the sun. Our weather is affected by the axis of the Earth in relation to the sun, it causes temperatures to fluctuate over the planet. The cycle of the birth, growth and death of flora and fauna are affected in all parts of the planet by the cyclical time periods of the world. As members of the human species we are also fauna, therefore, not exempt from the effects of this phenomena. We also know that the moon affects our planet on a cyclical basis and the pull of the gravitational effect of both the moon and the sun raises and lowers the tides in the oceans of the world. We know that we human beings are controlled by the cycles of night and day and that the spinning of our beautiful world controls our bio-rhythms. Half the world's population knows the cyclical effects of time and nature, because it has a direct effect upon the reproductive cycles of women. The ancients took this further and considered that birth and death were also cyclical and came to the conclusion that the destruction of the physical human form, or body, could not be the end of the story for an individual. This belief is still continued today in many major religions in the world and understanding this way of thinking is an important step in trying to understand our ancient ancestors. The concept with which they thought about death and rebirth is called reincarnation.

REINCARNATION

Reincarnation is the name given to the idea that there is no such thing as death of the soul. There is the natural time or cyclical running down of the material system known as the body. We all know this and most people try not to think about it or fail to come to terms with its inevitable truth. Put in modern terms, reincarnation could be more easily understood by those of us who use cars as an extension of our bodies with which to travel around in the environment more quickly and conveniently.

When we first buy the car it is new and efficient, but as time goes by it starts to deteriorate, the parts wear out and the car becomes more unreliable. Just like our bodies, it has a given life span or built in redundancy based on time. Eventually, when a car becomes too old to be serviceable, the driver gets out and changes the car for another. The ancients recognised a similar concept for the human body. One could say that they thought about the body as being a convenient vehicle for travelling around in the environment of our world, with the mind as the driver and the soul as the passenger. They saw the human being as body, personality, mind and soul. They knew that the body would die but believed that the soul would find another place, another vehicle, in the river of time in which to exist in the next material world. Here is what Wallace Budge says in his translation of The Egyptian The Book of the Dead:

> "*[1. According to the Egyptian belief man consisted of a body xa, a soul ba, an intelligence xu, and ka, The word ka means "image," the Greek ei? 'Dolon (compare Coptic kau Peyron, Lexicon, p. 61). The ka seems to have been the "ghost," as we should say, of a man, and it has been defined as his abstract personality, to which, after death, the Egyptians gave a material form. It was a subordinate part of the human being during life, but after death it became active; and to it the offerings brought to the tomb by the relatives of the dead were dedicated. It was believed that it returned to the body and had a share in its re-vivification. See Birch, Mémoire sur une patère Égyptienne (in Trans. Soc. Imp. des Antiquaires de France, 1858; Chabas, Papyrus Magique, pp. 28, 29; Maspero, Étude sur quelques peintures, p. 191 ff.; Trans. Soc. Bibl. Arch., vol. vi. p. 494 ff.; Brugsch, Aegyptologie, p. 181; Wiedemann, Religion der alien Aegypter, p. 126 f.)."*

The Kaballa may have been named after this esoteric idea and a literal translation could well be read as the Personality and Souls of God. Reincarnation had an important part to play in the reasons why the ancients kept time. The prophecy of reincarnation was predicted with the aid of astrology. A well known example of this combination of time keeping for the purposes of predicting the rebirth through the recurring cycles of certain types of individuals and events, is seen in the works of Nostradamus.[iv] Nostradamus used the ancients' system of predicting time through astrology. It is generally believed that this skill is applied through the study of astronomy, which requires the measurement of stars, the sun, moon and planets. Nostradamus would have used a quadrant and astrolabe (covered later in this book) for this purpose accompanied by almanacs, which predict the position of stars and planets throughout the year.

So, here again we see a division of two forms of observation where both astronomy and astrology have become separate from each other. One is now science and the other has become a belief and neither of them is compatible now, although astronomers still use the signs of the zodiac on their star charts. For instance, it is said that Sir Isaac Newton predicted the Second Coming of Christ in the year 1948. Those who have studied this say that he was wrong because there was no materialisation of Christ at this time. But were he correct, then Christ would have been born at that time and would be 53 years old now if Christ were an individual. Sir Isaac Newton's prediction was also likely to be based on astrology, which predicts birth dates under certain patterns and repeated cycles of the stars. In the original story of the birth of Jesus Christ, the three men of wisdom, from the east, who followed a star, predicted his birth not his recognition[v]. So here we have an example of astrology put to use in the bible.

Many religions in the world today still believe in re incarnation. To give some examples there are the Buddhist and Hindu faiths in Asia, and on the other side of the Pacific Ocean, the Aztec, Mayan and Inca peoples in the Americas also subscribe to this belief in reincarnation. The latter South American peoples, whilst embracing Christianity, still hold on to this belief, as do many spiritualist societies in Europe. This belief was also practised in Ancient Egypt and as I have already mentioned, their concept was that human beings had several parts to their entity and that the body is mortal, destined to die, but the soul is immortal and must find another body to inhabit after death. The Hindus pray to be set free of this wheel of life, for their understanding is, that they must return again and again so as to reach such a level of spiritual enlightenment, that they do not need to be born again and can remain in spirit form. They have perfected this to a degree of sophistication that a material caste system has been created in the society structure of Hindu India. The untouchables are at the bottom of the human pile and have to work their way up to higher castes through Karma and rebirth, by the work they do in this life. A cynic could say that the concept has been twisted so as to control the masses for the benefit of the upper classes. It must also be pointed out that the Bhagavad Gita[vi] also states that the untouchables are not the bottom of the tree of life, but there are Karmic offences that can take the soul further down the system, so that they are reincarnated as animal, vegetable or mineral. This concept is also taught in the Mayan Mystery Schools even today. Because of the cyclical forms of nature to be observed all around them, it was also natural that they should feel this way. The aim then was to become detached from the material world by achieving spiritual perfection. Buddhists considered the purpose of the enlightenment process as being extended beyond human form to the rocks, trees and all matter. That is why Buddhists are careful not to stand on an insect, as they consider the insect to be their brother and that all life is sacred. Buddhists also believe that some entities that reached perfection and could leave this Earthly stage, were of such high ideals and compassion, that their purpose and intent was to stay behind and help with the enlightenment of others struggling on the path. These entities would reincarnate at regular intervals to help with the spiritual growth of Mankind. There are also considered to be entities that are destructive to the spiritual growth of Man and that these also manifest or reincarnate at regular intervals. There is evidence of this line of belief as shown in the following quatrain giving the prediction, by Nostradamus, of the rise of Hitler and his association with the Pope Pius XII.

> *Dela partie de Mammer grand Pontife*
> *Subjugera les confins du Danube*
> *Chasser les criox par fer raffe ne riffe,*
> *Captifz, or, bagues plus de cent mille rubes.*
> *VI.49*

The following translation in English is worth reading.

> *The Great Pontiff by the warlike party*
> *Who will subjugate the borders of the Danube.*
> *The crooked cross pursued captives,*
> *Gold jewels, more than one hundred thousand rubies.*

This is an uncannily accurate prediction don't you think? This is a prediction of the growth of the Nazi party, which was formed in Austria, by the banks of the Danube. The Roman Catholic Church was accused of not standing up against the Nazis and even allowed them refuge in the Vatican after the war. There is also a disturbing intimation about the vast wealth that was stolen from victims by the Nazis during the Holocaust. But the most interesting coincidence is the mention of the crooked cross, which can only be a reference to the swastika. The true image of the swastika can be seen on the heraldic shield of the German Coat of Arms of the Masons.[vii] It is a depiction of the globe surrounded by four compasses or dividers located at the four cardinal points and opened at 90°. This symbol shows the concept of the measurement of the Earth. It is an ancient symbol that is now banned throughout the world, for the terrible associations that accompany its image. Yet, it is an image seen in ancient, pre-Columbian rock carvings from India, Egypt, Arizona and Mexico. It is becoming more understood by the general public that Hitler thought that the Aryan race was descended from the ancient "mythical" Atlanteans. It is also becoming more evident that he spent a great deal of time, resources and effort to discover more about ancient Atlantis and its fabled science. Astrology played a great part in Hitler's life. So, here is but one example of occult methods being used to predict the future reincarnation of players and events on the stage of life.

THE OCCULT

The word Occult is derived from a form of word meaning Hidden. An example of this in the modern world can be seen on the roof of police cars and ambulances. It is a type of light with a revolving mirror. The mirror spins mechanically around the light bulb first obscuring the light and then reflecting it toward the observer. It has the effect of being seen and then being hidden from view. The effect is to see a flashing light with a colour related to the glass surrounding the light source. This light is known as an occulting light

What I shall be revealing to you in this book may be the most important key to the understanding of the hidden cultures, magic and science of our ancient ancestors. In short, the findings in this book can be classed as "*occult*". You, the reader, may be the judge of whether or not this knowledge is still hidden and known to a chosen few in our modern society. Personally I believe that it is known only in fragments and spread between different religions, indigenous people and secret societies. I think that no one group holds the full knowledge and that division and secrecy keep it separated and divided. Alternatively, you may come to the conclusion that this is lost or forgotten knowledge rather than hidden. Nevertheless, to begin understanding these concepts and revelations requires an opening of the mind and an escape from the dogma and conditioning that we are used to. The effect of this revelation may result in an awakening of the mind and spirit of the reader. Most people will be astounded, surprised and confused by what they will read.

Reincarnation was also the belief of the Ancient Egyptians and the following excerpt from the Emerald Tablets, is more understandable having taken the concept of reincarnation and the occult on board.

THE EMERALD TABLETS OF THOTH*viii*

The Emerald tablets have long been the source of inspiration for seekers over the centuries and are believed to have their roots in Alchemy. The majority of people believe that Alchemy is the art of trying to turn lead into gold, whereas, the true translation would actually be the effort required in turning the base animal instinct of Man into a deeper spiritual being. The Egyptians were followers of this belief and illustrated it in many carvings, some of which can be seen in the

Louvre, Paris. There are stone statues in what appears to be varying stages of completion, some just showing a face emerging from a block of stone, others demonstrating the various levels of emergence, as though the mason had not finished the work. They were intended, in my opinion, to demonstrate the development of a soul as it emerges from the material state. The Perfect Soul shows a person standing on the block completely emerged.

Wisdom is the path to enlightenment. Wisdom is knowledge and knowledge is Power. Power corrupts if in the wrong hands and can be used for oppression and control of others less fortunate. That has been the way of the world for thousands of years. Conversely, put wisdom in the right hands and it may be used for the enlightenment of others.

The following is part of a translation of the words of Thoth, the ancient Egyptian God of Wisdom.

> *Nor for a time I descend, and the men of KHEM (Egypt) shall know me no more. But in a time yet unborn will I rise again, mighty and potent, requiring an accounting of those left behind me. Then beware, O men of KHEM, if ye have falsely betrayed my teaching, for I shall cast ye down from your high estate into the darkness of the caves from when ye came.*

> *Betray not my secrets to the men of the North or the men of the South lest my curse fall upon ye. Remember and heed my words, for surely will I return again and require of thee that which ye guard. Aye, even from beyond time and from beyond death shall I return, rewarding or punishing as ye have requited your trust.*

> *Great were my people in the ancient days, great beyond the conception of the little people now around me; knowing the wisdom of old, seeking far within the heart of infinity knowledge that belonged to Earth's youth.*

> *Wise were we with the wisdom of the Children of Light who dwelt among us. Strong we were with the power drawn from the eternal fire. And of all these, greatest among the children of men was my father, THOTME, keeper of the great temple,*

> *Raised I high over the entrance, a doorway, and a gateway leading down to Amenti. Few there would be with courage to dare it; few*

pass the portal to dark Amenti. Raised over the passage, I, a mighty pyramid, using the power that overcomes Earth force.

Deep and yet deeper place I a force-house or chamber; from it carved I a circular passage reaching almost to the great summit. There in the apex, set I the crystal, sending the ray into the "Time-Space," drawing the force from out of the ether, concentrating upon the Gateway to Amenti.

Other chambers I built and left vacant to all seeming, yet hidden within them are the keys to Amenti. He who in courage would dare the dark realms, let Him be purified first by long fasting. Lie in the sarcophagus of stone in my chamber. Then reveal I to him the great mysteries. Soon shall he follow to where I shall meet him, even in the darkness of Earth shall I meet him, I, Thoth, Lord of Wisdom, meet him and hold him and dwell with him always.

Builded I the Great Pyramid patterned after the pyramid of earth force, burning eternally so that it, too, might remain through the ages. In it, I builded my knowledge of "Magic-Science" so that I might be here when again I return from Amenti, Aye, while I sleep in the Halls of Amenti, my Soul roaming free will incarnate, dwell among men in this form or another. (Hermes, thrice-born.)

CHAPTER 2

THE GOLDEN THREAD

RESURRECTION

Resurrection is an entirely different matter and not to be confused with reincarnation. The Christian concept of resurrection is the hope given to Christians that the same body they lived in will be resurrected again after death. This seems to be based on the breaking of the cycle of reincarnation on the wheel of life, by the sacrifice of a Son of God for the original sins or karma of the people. In other words, an innocent and perfect person, carrying no inherited sin or karma, was the sacrificial lamb on behalf of the people. This innocent was known as Jesus Christ and by the evidence of his resurrection, the promise was given that continuous reincarnation in the wheel of life was not going to be a burden to the people who believed in him. Many believe he was the last of the Pharaohs and descended from the Davidic bloodline. Certainly, it could be construed that he was the last Pharaoh of the House of the Ram as he was declared the Shepherd of the Flock and he was incarnate at the end of the age of Aries.

It is indicated that this was the case following the story of the escape of his family into Egypt.[ix] It was from Egypt that he returned to Galilee and made his home in Nazareth becoming known as a Nazarene. It is told that Jesus became a carpenter, this was one who worked in wood and was involved in building. To be a carpenter was not a poor man's work as has been considered to be the case. Builders were held in same esteem as those who built Solomon's Temple and later the Mediaeval Cathedrals. These skills were based on religious principles, as considerable building work in those times was on religious construction and repair. They carried the respect of the church leaders as holy workers who practised their craft on religious buildings. Jesus and Joseph, his father, would have been members of a Guild, which would uphold considerable secrecy and furthermore, these members require the knowledge of geometry, mathematics and construction.

Christ was reputed to have spent considerable time in Egypt in Alexandria and Helipolis. Both places were the centres of learning for the Greeks after the conquest by Alexander the Great, several hundred years earlier. I will show you that the crucifixion was an insult far greater than that currently believed. The cross was an instrument for measuring and finding levels, for surveying, map making, astronomy, astrology and mathematics. Jesus Christ, the last Pharaoh of the House of the Ram, was brutally murdered upon the instrument of his own knowledge of science and spirituality.

The House of the Ram (Amen)

I will reveal to you a key to the deep and ancient knowledge left in the Great Pyramid of Khufu over four thousand years ago by the family of The House of Ram or Aries, otherwise known as the word used at the end of the Lord's prayer "*Amen*". Amen is the ancient name for the age of Aries, which started over four thousand years ago as the precessional cycle caused the sun at the spring equinox to leave the astrological age of Taurus and enter the age of Aries. The best known member of the family of The House of Ram is Tutankhamen. The meaning of this young Pharaoh's name is Tut meaning Thoth, Ahnk meaning life, and Amen meaning Aries. A literal translation would be Wisdom and Life of the Age of Aries. A question asked in a recent quiz show on British television, was: "*What is the meaning of Amen at the end of the Lord's prayer*"? The answer that they gave as the correct one was: "*So Say all of us*". This is the same answer that I was given by the local Church of Scotland minister when I was ten years old. These religious teachers are either misinformed or sadly mistaken, the congregation are being told to call upon the God of Aries (Amen) and to pray for the return of that Kingdom to come upon the Earth as it is in heaven. This is a misunderstanding of the nature of the cosmic and ancient clock devised by our ancestors. The real understanding is communicated to us by the Ancient Egyptians in the form of a phrase to help us understand this ancient clock. The phrase is "*As in Heaven so Below*". Aries is no longer the Kingdom of heaven as it was two thousand years ago. Pisces is the Kingdom of Heaven in present times and it will shortly pass away followed by the Age of Aquarius. To become close to God in the way that the ancients thought, was to be aware of your real place and moment in time. To do this you must have a key. That key was hidden by Thoth (Tut) the Wise, a mighty secret to be kept from the heathen, ignorant, destructive, materialistic children of the Earth, for millennia to come. A key that could only be revealed in times of freedom and enlightenment before the end of this age of Pisces. This is a key to a knowledge that was already ancient before it was used to build the pyramids in the ancient land of Khem (Egypt). Khem is the root for the word Chemistry and they were chemists and alchemists of the highest order. The Egyptian pyramids and temples were once gleaming beacons of civilisation in a time long lost to the world. They capture the imagination of modern people all over the world. Sir Isaac Newton visited them and so did Napoleon. Even today, in our highly advanced technical world, there is a mystery, mystique and deep awe for the observer viewing these massive edifices.

The pyramids of Giza are buildings of amazing technical accuracy and perfect beauty, that have been stripped of their glory and defiled in ignorance and greed over the ages, brought to crumbling ruin with their builders gone or lost long, long ago. But time fears the pyramids and they still hold their mighty secrets for us to discover. The knowledge is multi layered and starts with the simple ability to measure with geometry. It passes through design and construction, cartography and navigation, mathematics and astronomy, eventually to time keeping on a hourly, daily, weekly, monthly, yearly basis and onward and outward to the highest knowledge of the great cycles of the solar system and a total grasp of the realities of time.

This journey of discovery we are on, pulls us further into the hidden and ancient mists of the distant past, drawing back the occult curtains of how astrology, prediction and prophesy were achieved. We will explore the mysteries and wisdom of the ancient Gods of Time. We will try to fathom their way of thinking and see how it applies to us today. Let us go on to reveal the instrument left by Thoth for us to discover.

But who is Thoth?

Thoth, the Egyptian God of Wisdom.

The illustration at the beginning of this chapter depicts Thoth, the ancient Egyptian God of Wisdom. There is no one wearing the cloak and the cloak is universal and may be worn by one or many and that can be you or I. The Greeks called him Hermes, the Knights Templar called him Sophia (meaning Wisdom) or Baphomet who is represented by a skull in which there once dwelled a wise person.

Here is the translation of the description of Thoth from The Egyptian Book of the Dead by Wallace Budge:

> *"Thoth was the personification of intelligence. He was self-created and self-existent, and was the "heart of Ra." He invented writing, letters, the arts and sciences, and he was skilled in astronomy and mathematics. Among his many titles are "lord of Law," "maker of Law," and "begetter of Law." He justified Osiris against his enemies, and he wrote the story of the fight between Horus, the son of Osiris, and Set. As "lord of Law" he presides over the trial of the heart of the dead, and, as being the justifier of the god Osiris against his enemies, he is represented in funereal scenes as the justifier also of the dead before Osiris (see Lanzone, Dizionario, p. 1264 ff., and tav. cccciv. No. I; Pierret, Panthéon, pp. 10-14; and Brugsch,*

*Religion und Mythologie, p. 439 ff.). Brugsch connects the name
Tehuti (Thoth) with the old Egyptian word tehu, "ibis," and he
believes that it means the "being who is like an ibis." The word tex
also means, "to measure," "to compute," "to weigh"; and as this
god is called "the counter of the heavens and the stars, and of all
that therein is," the connection of the name Thoth with tex is evident.
Bronze and faïence figures of the god represent him with the head of
an ibis, and holding an utchat in his hands (see Nos. 481, 490a, and
11,385 in the British Museum). The utchat, or eye of the sun, has reference to the belief that Thoth brought back each morning the light
of the sun which had been removed during the night".*

So, what is it that Thoth reveals to us from the depths of the Great Pyramid of
Khufu?

This incarnation of wisdom in one, many or all. The balance of Truth and
justice, mathematics, computation, the counter of the heaven and stars and all
that is within them. The ability to weigh and measure. How perverse is it that
he was seen as an Ibis by the ancient Egyptians and that the son of Prince
Charles and Princess Diana of the House of Windsor recently shot one by
accident. What kind of omen could be read into that accident by an ancient
Shaman?

Thoth reveals to us an instrument that has not been understood or noticed in the
twentieth century by laymen and academics alike. An instrument that can carry
out all the functions considered as being the skills of Thoth, the God of Wisdom.
That instrument of knowledge and practical skill is the cross. Not the icon of
the cross or the wooden ones, or those of stone on church steeples, or on altars,
or the graves of the dead. Nor is it those images depicted in paintings and
drawings or any other form of literature or the ones worn by religious leaders or
by ordinary faithful people that are made in gold, silver or pewter. The instrument of Thoth is not to be found among those crosses that are encrusted with
diamond and other precious jewels or sung about in songs throughout the world.
The real cross, the working cross, is a living, functioning practical instrument of
measurement that can be made and used by anyone. Let me introduce you to
this superb instrument of the ancients, this bridge of understanding and knowledge as I have found it to be. When you see the beauty of its form and function,
you will no longer be mystified by the mysteries of the past and the achievements of our ancestors. You will no longer read books on the ancient past and
wonder, as the authors themselves do, how the construction and use of ancient
monuments were achieved and why. You will wonder, as I do, why this has not
become common knowledge in our time.

THE CROSS

What do you see in the photograph below?

What does the cross and circle within mean to you? The twisted decoration in the circle and the motifs carved on the surface? Do you understand that no architectural object, small or large, was made in the past by religious craftsmen without deep meaning, both hidden and obvious? Is that one of the reasons why Prince Charles of the House of Windsor criticised modern architecture for lacking spiritual depth? The cross you see above is shown with a wheel like a cart, it is on a hub like any other wheel that is meant to spin freely. I will show you that it is a plan made in stone for an ancient instrument where the wheel does indeed spin freely, and the action of this spinning wheel allows this instrument to measure all sorts of things. Why then, is this cross and many forms like it depicted on churches and gravestones all over the world? It is mostly seen in connection with death and spirituality. Why?

The photograph depicts a Celtic style cross and hidden in the carvings of the cross are messages about the path to enlightenment as seen by our ancestors. They believed that enlightenment could only be achieved through personal, deep awareness of self and the oneness with all of the world, the cosmos, nature and the cycles of time. With knowledge of how to use the cross in a scientific and practical way, they were able to achieve a high science that is lost to us today. There are many keys to these ancient mysteries. The keys are kept and guarded, hidden and yet obvious, left for us to discover by our wise ancestors down the ages. For those who are seeking the meaning of life, the questions about why we are here and what happens to us when we are no longer here, the cross, as I shall demonstrate, is one of the most important keys in revealing the way our ancestors thought.

Let me assure you that what you are looking at is not what it seems on the surface. Some have always considered that the Celtic cross is a religious symbol that represents the sun and the four cardinal points. Constantine the Great even intimated that this was the case and although the explanation seems to make sense, this is not actually so. Let me show you that this object is a practical working tool of great depth that is embellished with esoteric symbols connected with time, life, death and spirituality Many scholars assign the carvings on the surface of the cross esoteric meaning, without really understanding the concepts and beliefs behind them. The Vine of Life can be seen, as can the intertwined Serpent. You may have your own interpretations of what these illustrations represent but generally they mean order and chaos, darkness and light, Ying and Yang, Nature and the cyclical forms of Time. The symbol of the cross is another message in itself, written in stone for those with eyes to see. Many messages were left in stone and in myth by our ancestors waiting to be decoded at a later date. One of the greatest of these messages I believe, being the Great Pyramid of Khufu at Giza in Egypt.

As I have already told you, this pyramid was built at the start of the age of Aries. The ruling family of this age known as Amen, the Ram, built it to celebrate their deep understanding of time and the Creator. Some of the names of the Pharaohs were Amenhotep 1, Amenhotep 2, and the most famous that you will have heard of as I have already said, is the name of Tutankhamen. The name has become twisted to cause confusion and is frequently translated as Amun.

This great pyramid and other pyramids were used as multi functional clocks and calendars. In the mighty Pyramid of Khufu is a secret, one of the keys of Thoth (Wisdom). Hidden deep within its chambers, is the remains of a cross. Parts of this cross were found in 1872, but unfortunately no one recognised the historical and religious significance of these remarkable items until 1998. To me, this missed opportunity of discovery was and still is one of the most important unrecognised archaeological finds of all time. I am honoured to have been the first person in 4,500 years to recognise it and its abilities. This cross was left by the architect of the Family of the Ram to awaken the memories in those reincarnated souls as they returned to the material world and to lead them to discover the mysteries of the concept of time, mathematics and measure. There is more to the cross than just being seen as religious iconography, it is an instrument. An instrument that was of vital importance in the building of the Great Pyramid, thus proving that the Cross is older than the Great Pyramid. Conversely, the Great Pyramid could not exist without the Cross.

The Cross is an ancient working, scientific instrument of such antiquity that the knowledge of its existence and its use have been lost in the mists of time. The authorities in the time of Christ crucified him upon its image either as an insult to his knowledge or as a warning to all others not to tamper with the most powerful tool of its time. Crucifixion as a form of execution for Jewish people was a rare occurrence in Roman held Jerusalem, since Jewish people were generally stoned to death for capital offences, whereas Roman offenders were crucified.

THE INSIGHT

I have written this book in the hope of revealing the merits of this amazing instrument to the searcher, to those seekers looking for truth and meaning to history and life. The writings and revelations contained in these pages will reveal many of the lost and forgotten scientific methods employed by our ancient ancestors. Our ancient forefathers strove to gain an insight into the workings of the cosmos and they gained a knowledge and a way of thinking that surpasses our own in many ways, despite all our modern day advancements.

Most of what you will read has never been revealed to the uninitiated before. These ancient occult techniques are the keys to the understanding of our past, our present and of our future. How did the Mayan Calendar prophesy certain events such as the destruction of the northern hemisphere by fire in 2012? This date also happens to be the next time that the sun reverses its polarity. Is this co-incidence? The last reversal, according to some scientists, happened in 2000 and it happens every 12 years without fail. How could the ancient Mayan priests be aware of this? Modern day science is only just discovering that this is the case and is only able to understand this by measuring the sun with x-rays taken by the space probes that NASA sends into space specifically for this purpose.

The extreme weather associated with severe sunspot activity is increasing every 12 years as the sun reaches its solar maximum. During this cosmic event, the mass population is blamed for this potentially fatal occurrence as man-made CO_2 emissions are considered to be the culprit for the rising temperatures. This is not so and the public is being misled as they have been for thousands of years. Remember that the concept of service is *"A government for the people by the people"*? Rather than what we have which is *"A government of the people"* and falls far short of the ideals of our ancestors. Where is the wisdom of Thoth in that? We are told only that which allows us to be controlled. The global warming that is occurring and threatens to raise the sea levels, is part of a cycle that repeats over and over again on our planet and our ancestors seemed to be aware of it and prophesied it.

Within these pages are some of the most disturbing and surprising revelations ever made to modern man. The knowledge that has been bestowed upon me is considered by me to be a gift. A gift that was discovered due to the archetype being awakened in me as I undertook an in depth study of the pyramids in Egypt. I believe it is my duty as the author of this book, to share this knowledge and revelation with as many people as possible in the hope that they will listen and understand. There is something for everyone and you, the reader, will take from this publication, that which is important only to you. Some may gain greater spiritual understanding of our place in the universe. Others may just be fascinated by the simple beauty of the methods by which the mysteries of our ancestors have been solved. Practical people such as astronomers, mathematicians, navigators, and surveyors will be intrigued. Scientists and archaeologists will be assisted in understanding the very roots of our culture and science. By reading and digesting the findings of this book, some will find it easier to teach practical subjects such as mathematics, history, surveying, map-making, navigation and astronomy. Most importantly though, the children who will be taught about these findings, will understand in practical and common sense ways, how to understand the theory. It will be simple for them to decipher the mysteries of time and calculate their positioning on the planet upon which they, and we, live. I believe that my discovery is more for the children of our future than for us. The as yet unborn, may need this knowledge to survive in our planet's uncertain future.

This forgotten knowledge was the basis upon which the civilisation in which we live our life today was formed. These are the fundamental roots that lead to the understanding of our modern day religions, technology, mathematics and the structure of our society as it presently manifests itself. These are exciting times and much is owed to the development of information technology and electronic communications. This advance in cross continental communication allows the instant and free exchange of ideas and knowledge that was denied to those who came before us. Researching and accessing knowledge is at a height unprecedented in history. Censorship by those who seek to control the masses, so as to profit by their ignorance is difficult to implement in this modern time. Freedom of knowledge enables this holiday from ignorance and suppression. However, as technological methods advance, certain authorities will seek to control the Internet, and this valuable source of communication may come to an end thus enforcing a limitation on the amount of knowledge able to be gained, thereby imposing upon us a certain level of censorship.

THE ONENESS

I will attempt to show you that we, as individual human beings, are also part of the whole or One as the ancients saw it. An example of this way of thinking is best described as a comparing the individual to the cells in a body. One cell is part of the eye and its function is to allow the whole organism to see. Another is designed as part of the ear and its purpose is hearing. Millions of cells make up the complete human form with their genetic instructions, gifts and attributes, but although each cell is an individual and many are born and die throughout the life of the organism, they are all part of the oneness of the form they maintain. This is so with individual humans as they maintain the current form of society and mankind as a whole. Furthermore, this is the purpose of all forms of life. They are interconnected and all have a purpose that maintains the balance of life on the planet. There are no useless creatures, they all keep the balance for the good of the whole.

In the depiction of Thoth, the rod and staff are the all important elements as together, they form an instrument that, when coupled with ancient wisdom, becomes the very basis of all the teachings passed down to us in hidden form through the aeons. It is the wisdom that may be read in the many religious books, both ancient and new. Clues to this wisdom may be found in the Bible, Koran, Dead Sea Scrolls, Torah and Baghavad Gita. Other clues can be found carved in rock all over the world. Clues that are in artwork, myth and poem alike. It is the wisdom of the cycles. The Celtic cross represents the oneness by showing the cosmic cycles within its wheels and the oneness of male and female. The arms reach out toward the cosmos and its centre is deep within the Earth. It is a bridge between the practical and earthy to the spiritual and cosmic as are our minds. It is only in this time in which we live, that it is possible to learn of these things. These have to be seen as wonderful times as they are, I believe, the promised times of enlightenment and the window will not be open long. This is not a new enlightenment for mankind but a remembering of what we once knew, it is a lesson that is to be learned from the ancient knowledge and wisdom of the past that could be re-integrated into our modern way of living. Enlightenment can already be seen in the revolutions in thinking that have changed our society in the last few years. People have been seeking all sorts of answers through music, art and politics. Many are trying consciously and sub-consciously to seek a new and better way that is in harmony with the Earth and nature and at odds with material values that rape and pillage both the planet and the people.

We live in a dysfunctional society that is headed for its own social, material and spiritual destruction if it does not become aware quickly. The creation of the present society began thousands of years ago with persecution of millions and the deliberate destruction of knowledge that was in conflict with new beliefs. For over two thousand years, free thinking people have been horrendously tortured, betrayed and destroyed by those who thought it in their interests to suppress anything that would be considered a threat to their control over the masses of mankind. How naive the perpetrators of these crimes and suppression were. Do they not know that only a small percentage of Humanity will be interested in looking for the answers and even less will care. Most people are too busy with the slavery of earning a living to have time to question their lot, let alone their leaders.

In third world countries, the populations are kept in war, starvation and disease to such an appalling level that they are no threat to the ones in political or religious power. These secrets held by the perpetrators of crimes against the Soul of Humanity were always safe, the horrors they inflicted on the few dissenters and seekers was needless. The majority of mankind is content to be guided from life to death by others, *"Died at twenty, buried at 60"* is a saying that is well worn and well proved.

You, the reader, are different, or you would not have picked this book up and decided to read it. You have taken a chance to discover the truth for yourself and on this path you will see things that will surprise you. You will see things that have been with you and part of your surroundings all your life and yet you have not previously understood them. To read of the things that are in this book in earlier times, would have resulted in your persecution and maybe, even death.

Do you think history and religion are exactly as taught to you by the priest and the teacher? Remember the rock song written by Pink Floyd? The words went, *"Teacher, teacher, leave our kids alone, don't put another brick in the wall"*. Do you feel that we each have a wall that does not allow us to see the world as it is? Are you aware of the modern Zen Buddhist saying, *"Life is like a glass coffee table, covered with rubbish, clear it so that you can see through"*? To pursue this path means awakening the Archetype. Do you think that this applies to you? Do you want to challenge the way you think, be changed for ever? If you do not, then close the book now and put it back on the shelf for if you continue to read this your mind may waken and will become full of more and more questions. Questions that may shake the very foundations of your life and beliefs. You will need "Faith" and strength of mind to seek answers to these questions for you may no longer be a sheep in the flock or a bee in the hive. You may gain sovereignty over your person. You will see things written and illustrated here, that will make you ask "what was wrong with the state of my mind that I could not see this before?" What is the condition of everyone else's mind that they are blind to these things also? This is not a question of intelligence or privilege, but more a question of freedom, of being open minded and curious. This is what Christ meant when he said *"Be as Children"*.

Thousands of politicians, scientists, historians, theologians and archaeologists can not see these things which I have found and have not been able to see them for countless generations. It gets worse though, for there are amongst these groups of people, some of whom are in positions of great power and influence, who do not want their empires shattered by changes and revelations that may well undermine the work of a lifetime. They are supposed to be seekers after truth and yet they sell their youthful ideals for "material" gain or for maintenance of the status quo. They trade real gold for fools gold and the glitter of the illusion of fame and power. They are the worst hypocrites of all for they deceive the innocent and deprive them of the truth. There are those who do see but are also very afraid of being the first to speak in case they are wrong and are ridiculed in front of their peer groups. There are those who are the self appointed protectors of the "*inferior mentality*" of the common man. How dare they deny people the chance to grow or fail to give them due credit for being able to discriminate falsehood from truth. You may also ask the question, are there those who already know the things that are revealed in this book? Is there a conspiracy to hide these things away from us by some of those who are in power today? Let us explore the nature of time as the ancients saw it and see if there is any reality in this view with the world as we know it today.

THE GODS OF TIME

I call this section the Gods of Time in acknowledgement that the keepers of time were divine beings to the ancients. Time is the most important element in understanding the workings of the Earth, cosmos and life. We all take time for granted, yet our allotted life span on this Earth is wholly governed by time. Our life on Earth is dominated by time, for example we wear a watch to tell us when to be at appointments, when to go to bed and when to get up and our personal body-clock only stops when it is time for us to leave this Earth. However, the watch is a relatively new invention and has only been perfected in the last few hundred years. So, how was time kept before this invention? Unfortunately, we have only a very hazy understanding of how this was possible in ancient times, the methods having slipped away into obscurity.

TIME

So how do you understand time? Most people have, at one time or another, enjoyed the scientific fiction of time travel. H G Wells work and fiction called The Time Machine is one such example and Arthur C Clarke has written several books on Time and its effects. Space Time occupies scientists all over the world, they look at quantum mechanics, physics and worm holes to find the answer. If people really understood the concept of time, all wars and conflict would end. Wars, conflict, grudges, revenge and discrimination are caused by memory and the hearsay of memories only and any policeman or judge will tell you how faulty that ability is in most humans. Think of the horrors that memory has inflicted on the masses of humanity. Hate of families, countries, religions and individuals for crimes perceived to have been carried out by ancestors in the distant past. This is an illusion for there is no past, it does not exist. Science tells us that nature does not waste energy and that we live in a dynamic changing universe.

THE PAST

To try a personal experiment with time and the past, look at a star in the night sky and remember that you are not seeing the star as it is now. You see it only through the photons that reach your amazing eye and are caught there for an instant. The cells that make up the mirror at the back of your eye are designed to catch photons and transmit the information to your brain. *"It is not so much the star that is a miracle but the eye that perceives it."* The star photons may have been travelling at the speed of light for thousands and even millions of years before you alone caught them. Here is the illusion, the star may not even exist now, it certainly is not in the same position in the heavens as it was when the photon you caught in your eye, left on its long journey to Earth. It is in the past.

Try a closer experiment next. When you look at the moon and catch the reflected photons from our sun as they bounce back to us through geometrical reflection the moon has moved fractionally from the position that you view it. The moon you see is in the past. The same exercise applies to a street lamp, what you see is in the past.

When you talk to a friend and he or she replies and your ear catches the sound waves, they are fractionally in the past as is your vision of your friend. Sound and light travel at different speeds and your senses create the illusion that what you see and hear is in the present, but it is not. It is illusion.

Touch, is that in the present? Unfortunately not. If you touch your toe or even your nose with your own finger, what you see is in the past and also what you feel. Feeling is transmitted by electrical and chemical relays to the brain and takes time to reach the processing centre. It is fractional in time delay but never the less, in the past and illusion. So where is your consciousness seated?

How small is that place in time in which *you* exist. How narrow is that moment between the past and the future? Where is the real you to be found? It is not in the sensations of your feet, finger or eyes as they are in the past and do not exist. Are you then alone? Am I? Or are our minds the only part of us that is existing in that instant of reality called the moment of now together? The words that you read in this book are already from the past and I am no longer sitting at the keyboard typing. Let us try another experiment to help make us aware of the nature of time as it applies to us. Try snapping your fingers and then reaching back in time to stop it happening, you can not because although it happened, it no longer exists. Do you think that there is an image of you back in the past forever snapping your fingers? I am afraid not as there can be no past that is in existence.

We know this because science and logic establish that nature does not allow the waste of energy, none is lost only changed. We are a discriminating species, we need to discriminate to survive, so we have difficulty seeing and understanding this fundamental knowledge and consequently live in a state of illusion all our lives. When we drive along a road in our car, we imagine the road we have travelled to still exist and yet it does not, it is a different road, stones have moved and it is more worn, it may be wet or dry and the wind that travels over it is different to the one that blew as we travelled by. The road we remember as we travel in cars and life does not exist after we have passed on our way. The ancients knew this and therefore also believed that the dead were not in the past but in the present. Not dead, but changed. Not gone, but here. But then, the logic of that way of thinking extends beyond the personal and says that all things that ever were are here and now. Even the dinosaurs that lived millions of years ago are changed into another form of energy that still exists here and now. The ancient wise ones also believed that the cosmic mind remembers all things and that there is nothing new under the sun. To understand and discover these memories meant achieving the seemingly difficult path of finding a way to tap into the cosmic mind. An example of this link between man's discoveries and the ancient memories of nature is the discovery of the box girder, this revolutionary way of strengthening buildings which made possible the construction of skyscrapers. The box girder was considered to be unique until it was found that nature had already beaten us to the punch. The flying dinosaur known as the Pterodactyl was quite large and had a lot of weight to carry when it flew, recently it was discovered from fossil records that its wing bones were made of a type of box girder construction that nature invented millions of years ago.

THE FUTURE IS ONLY POTENTIAL

By the same logic then, if the past does not exist, then neither does the future. The narrow band of energy between the two voids of nothingness in which all things exist is so small as to be immeasurable. Potential exists within the present and that potential can affect the future. The ultimate expression of this potential can be seen in the design of a seed or a human cell. Written in this tiny package is all the information required to make a full grown individual capable of producing its own seed. It is the only way we travel through time. Farmers know this and plant seeds for harvest later on so as to feed themselves and their families. Only cause and effect can influence the future. "You reap what you sow."

This was described by the ancient wise ones of the east as "*Karma*" or by the Bible as "*Original Sin*". Karma is the result of living in such a narrow band between the past and the future; "The Now". Because the Now is so small, an individual has no time to think in a sudden situation and therefore reacts to the way in which he or she has conditioned themselves. Ego is the cause of this problem and is largely materially based. "The seven deadly sins" are reported by the ancients as such dangerous manifestations of "*Ego*". Fear is the greatest of the ego or self centred reactions and the individual often regrets their reaction afterwards, particularly if it produces a bad effect that is dangerous to that individual later, or if they are higher thinking, to their family or society. So, the Ancients realised that the individual self was a small part of the whole or One as they called everything that was in the present. They believed that fertility was the only way to make sure that they were able to be immortal on the planet and the Ancient Egyptians made a great deal about this part of life and existence. This was to be seen in all their temples and phallic symbols were to be seen in profusion at the height of their power. You see, they could not be reincarnated as a human, if there was no new and suitable form for their soul or Ba to inhabit in the future. The early Christians destroyed the evidence of these phallic and fertility carvings and other representations of sex in the raw because they were offended by them. Even the Victorian English were unhappy about such things and covered these sights in the museums of the day. The Egyptian concept of the importance of fertility was extended to the growing of crops for the feeding of the people and livestock. They were acutely aware of the cycles of the year and measured the time of the cycles of planting, harvesting and storing. They created only three seasons in the year instead of our four and these were based on the annual inundation of the Nile as the rains in the mountains around Lake Victoria some 3000 miles away released floods that swept waters carrying silt rich in nutrients and minerals down the river, depositing them on the banks of the Nile. This event needed to be predicted so as to prepare the planting. Things were different then than they are now. They had no clock as we know it. The only way to predict the season of the inundation was by the stars and that is exactly what they did. They used the great clock of the world by measuring its rotation against the background of fixed stars at night and the sun during the day. To do this they needed an instrument and that instrument was the cross.

The cross was not invented by the Egyptians, but inherited, hidden and subsequently lost. This loss resulted in their inability to keep time accurately and we have the same problems today. The Romans and others tried to correct this with only part of the knowledge and the result is several different calendars throughout the world that are in conflict with each other and have an inability to keep pace with the moving forward of the seasons caused by Precession of the Equinoxes. The first clue to the ancients' depth of knowledge comes from understanding that they were aware of time in the method that I have described above. That all creatures and entities live in the Now. The moment of now that is a time zone between No Thing and No Thing. The abyss of nothing.

THE TIME ZONE

The enigmatic face of the Great Sphinx stares east at the equinoctial point of the rising sun on the eastern horizon. This vast sculpture has been there longer than the memory of man, even longer than our written or recorded history and is a huge cause of consternation in the world of academia. Various historians and seekers of the truth have tried to date this carving of limestone built on ancient bedrock. They are as obsessed today with trying to date it as they were a hundred years ago and so still the controversy rages on.

There are researchers and theorists who sincerely believe that the Sphinx is older than the date of 10,500 BC and was built before the great flood, and then there are also those who believe it to be a construction left over from an advanced civilisation long destroyed by a cataclysm of such proportions as to be unimaginable. Others are of the mindset that it was built in the 4th Dynasty of the Egyptian Pharaohs and that it represents the face of a long dead king and further still, there are also those who consider the Sphinx to be a representation of the constellation Leo. Who is right?

The ancient Arabic name for the Sphinx is *Horem el Akhet*, literally translated by some to mean *"Time on the Horizon"*. It was also known as *"The Terror"* or *"The Strangler"*. The Sphinx is also a representation of life and death and is in itself linked to "Time". To understand what was meant by that, one has to leave the safety of the cushioned world in which we live, with our concepts of materialism and all that goes with it. We have to try and understand why death was an all consuming thought to the ancient Egyptians. These people were not privy to the modern luxuries of television, motor cars and air conditioning. They were not party to modern medicines and dentists. They were mortal and were extremely aware of this fact. Death and suffering surrounded them on a daily basis and they looked to salvation and a better life after death. They would look forward to rebirth as a star. As we are not of the same mindset as the ancients, we have to decipher all the available evidence for a brief insight as to what was occurring in those times, so let us examine the meaning of the words.

Today we take so much for granted. The things we see, even the words that we speak. We learn these words as children by listening to our parents and witnessing their use in daily life and we begin to use them without giving their origin or meaning a thought. For example, *In the beginning was the "word"* according to *Genesis* in the Bible. The *word* is the manifestation of intention, making thought a reality through the power of communication. Likewise the word for *magic* is *spell*, this also being the word used to describe the placing of the consonants and vowels in their correct order to create these words e.g. "*spell that word correctly*". I am casting a spell with the writing of this book. I am putting ideas into words so that others might understand what I am trying to say. When the spell or seeds of thought that I am casting fall on fertile soil they will grow and spread until they become part of the mind of Man.

The same is so of the word *Horizon*. To understand the meaning, you have to break the word down, the first part being *Hor*. The root of this section of the word is derived from the ancient Egyptian myth of Horus and his followers, "*Horus*" was the Egyptian God of Time. He was seen as a falcon and from this root the word "*horology*" is derived, meaning the science of time keeping. The second part of the word is "*zon*" meaning zone or area. So, we have "*TIME ZONE*" as an understandable result.

The Great Pyramid is also known as "*Khufu's Horizon*". This name could represent Khufu's allotted life span on Earth as a man manifest as God incarnate, guardian against chaos, maintainer of the Sun, maintainer of the annual inundation of the Nile and the regular planting and harvesting cycles of life. This ancient knowledge of life and death is reflected even today by the secret Masonic question to find out if another is a brother Mason.

> Question: "*Are you a travelling man?*"
> Answer: "*From the East to the West*"

Where does this originate as an idea in the Masonic brotherhood? It is knowledge passed from The Knights Templar to the Masons from their discoveries of Ancient Egyptian wisdom that was hidden in the Temple of Solomon long before the destruction of Alexandria. It is the fundamental starting point for timekeeping, measuring and travelling the planet on which we live. Here is a passage from The Egyptian Book of the Dead, in which we find the following clue:

> *6 Heru-khuti, i.e., "Horus of the two horizons," the Harmachis of the Greeks, is the day-sun from his rising in the eastern horizon to his setting in the western horizon; for the various forms in which he is represented, see Lanzone, Dizionario, tav. 129. Strictly speaking, he is the rising sun, and is one of the most important forms of Horus. As god of mid-day and evening he is called Ra-Harmachis and Tmu-Harmachis respectively. The sphinx at Gizeh was dedicated to him.]*

East meant life and the rising of the giver of life, the Sun. West means death as the setting place of the Sun as it entered the nether world. All living entities have horizons or time zones in which they live out their allotted life span. The ancient Egyptian Sphinx stares at the Eastern Horizon looking for the place of the rising sun at the equinox. It forever watches the start of the new day and the new cycle, as life renews each morning from the first dim golden glow in the eastern horizon. It sees blindly through limestone eyes, closed against the sharp abrasive desert sands, around seventy spring equinoxes in the life of a man. It ticks away the minutes, hours, days and years of every mortal creature on the planet, including you and I as we travel "*from east to west*" like the sun.

There are those who say that there are vague records to be found showing that there was a twin to the sphinx and that it faced west to the point of the setting sun. There is as yet no proof of this, but it would certainly make logical sense. This book is about time and our ancient ancestors' concepts of this phenomenon. It is not concerned only with their understanding but also with the methods they used to capture time and were able to turn it into a science far beyond our current understanding. A science and understanding that beggars belief. A science that goes far beyond the practical and worldly as we currently know it. Their lost knowledge was so great that they were able to construct pyramids that were both incredibly accurate and vast. Buildings constructed with apparently limited technology, a feat that is unrepeatable today even with our more advanced tools. It was a feat that enables us to refer to the Great Pyramid of Khufu at Giza, as one of the seven ancient wonders of the world. This one pyramid weighs an estimated 6,000,000 tons and appears to be built on a superior understanding of mathematics and astronomy, far beyond the capability of the current comprehension and knowledge that we give our ancient ancestors credit for. But more so, it shows a motivation greater than anything we achieve as a team today.

Some try to say that it was built by slaves, but it was not, it was built with the kind of love and devotion that can move mountains. Musicians find inspiration from the pyramid based on harmonics. Geometers find correlations that are astounding and encompass all the geometrical shapes possible, including the fibonnachi spirals and platonic solids. Mathematicians argue about the right angled triangles and the use of Pi. It is a source of constant wonderment to all those who encounter it. The knowledge surrounding this vast precision instrument that the Greeks called the first wonder of the ancient world has long been struggled over. The knowledge taken from Egypt over the millennia has been suppressed and hidden, often with good reason as much of it has been deemed witchcraft, demonic and satanic. It may be so, I am not judging, only reporting, you may make up your own mind. This hidden knowledge is known as "*The Occult*" and the ability to use it is known as having second sight or the third eye.

CHAPTER 3

THE
THIRD EYE

THE THIRD EYE OR SECOND SIGHT

I believe that the discoveries revealed in this book may help in the resurrection of the lost memories of our ancient ancestors in some people. These are ancestors who are still with us today in the view of both the ancients and some modern societies. The ability to see the occult or the "awakening" is what the ancients called, "*the third eye*". We all carry inherited memories that are hidden deep within our genetic makeup. The majority of us are totally unaware of the existence of these subconscious memories and abilities, so very few of us are able to understand the significance of them. The Scottish Masons and Knight Templars were aware of this second sight and said so. Henry Adamson of Perth, Scotland stated in his "*Muses Threnodie*"[x] in 1636:

> "*For we are Brethern of the Rosie Cross,*
> *we have the Mason's Word and second sight*"

There are buildings and artefacts that were designed by Masons to awaken the archetype housed deep within the subconscious. Jim Bowles has already explained the concept of Archetype in the introduction to this book in his own way. The archetype is the basic spiritual or practical truth from which all hybrids and intricacies are evolved. The "*Celtic cross*" is the archetype of the sextant and the theodolite. It is also the archetype of the quadrant and the astrolabe. It is the physical manifestation of the ancient instrument behind the archetype of all construction, surveying, mathematics, astrology and astronomy. In the Bible, the archetype of Man is known as Adam and the archetype for Woman is Eve. The complete soul or Super Soul was known as Atum or Atman in ancient Egypt. In the Hindu faith the Supersoul was known as Krsna or Krishna and the following description can be read in the Bhagavad Gita.

> "*Earth, Water, Fire, Air, Ether, Mind, Intelligence, and False Ego-*
> *all together these eight comprise the separated material energies of*
> *the Supersoul*". [xi]

It was believed that the archetype of mankind's soul or spirit, as a complete entity, might be reached by rising above the illusions of the material world in which we live. "*There is nothing new under the sun*", all creation is in the cosmic memory so to tune in, just sit and be still. This is achieved by sitting quietly and emptying the mind of words and thought. A state of mental silence and stillness is very difficult to achieve for the unpractised as most modern people can not stand silence for long, however all is not lost as people can also achieve this state of awareness when they dream. Dreaming listens to the cosmic mind. Scientists do not understand dreams even today and are still carrying out intensive research in this area. To explore the subconscious and the cosmic mind, consider this question for a moment. How is it possible to frighten oneself in a dream?

If we are separate individuals living solely in our own minds, then we must be responsible for creating our own thoughts. If we create our own thoughts, then how can we envisage something new, funny or frightening in our dream? How can we possibly surprise ourselves with something we already know? It's like laughing at our own jokes. The logical conclusion is that we don't know it, but our subconscious does and so it can surprise, frighten, amuse and teach us. This is the level to which the ancients constantly strove to achieve a oneness with the cosmos. Their dearest wish was to be in touch or at one with the Supersoul. It is my belief that this is one of the reasons why our ancestors put such buildings and designs in these specific places. Take cathedrals, for instance, most of the architecture in these structures is designed to reflect the cosmic concepts in the mind of the observer by incorporating the mathematics, geometry and harmony of the universe. The idea is to awaken the archetype in the reincarnated individual and to remind them of the spiritual journey they are on.

For a lot of those people who have already had a preview on my website of the ancients instrument of wisdom, their first reaction has generally been one of an immediate recognition of truth, closely followed by excitement and a feeling of intense curiosity. They feel as though they knew about this revelation already. After the initial surprise has passed, it is the usual response from many that some pertinent questions need to be asked: What is happening to our current understanding of history and religion? Why has no one recognised this common place object before? Out of the millions of people who have observed the Celtic cross over thousands of years, why has this revered and sacred icon not been revealed for what it really is before now?

If we are not able to understand something when it is all around us, what else do we miss and misinterpret. It is common knowledge that the world is not always as we see it, but why is this so? What conditioning of our minds and what false cultural beliefs have been instilled into us by our parents, society, teachers, religious leaders, politicians and leaders of all sorts, over the millennia? Who are we really? Where do we come from? Where are we going?

To begin to discover those answers we need to develop our third eye and ask questions.

Was there once a highly advanced civilisation in the mists of time that was wiped out by a great flood and cataclysm? Could this have been a civilisation that was so fundamentally different from the one we have now, that we are unable to recognise it? Or is current academic belief right in the assertion that we are a progressing species on our way to greater development through natural selection and evolution? Could they be wrong in this linear thought pattern? Does the increasing sophistication of our technology and great advances in medicine prove us to be more intelligent? No, there is no real evidence to suggest that we are any more intelligent than our ancient forefathers were, but there is considerable evidence that we have forgotten much of their knowledge. There is no doubt that the division between religion and science is growing and that art is becoming secondary to scientific discovery, this is understandable after the persecution of science by religion over the millennia. The Vatican has only just recognised that the Earth is not the centre of the Universe and only recently accepted Darwin. The Pope is on a mission of reconciliation and apologising to the peoples and nations of the world for the terrible and ignorant actions of the Church in the past. His burden is greater than can be imagined and a compassionate person can see its terrible weight upon him.

Diabolical is a good word for evil as it means divided. Division is the problem and division weakens us. Division causes loss of knowledge in the past and the present as the latest evangelistic society tries to obliterate all the knowledge of its vanquished foes. The Romans, Christians and Moslems were guilty of the offence of burning books and destroying knowledge in Egypt and Alexandria. The Romans obliterated the ancient ways in Britain. The Christians, under the banner of the Conquistador and with the cross as a symbol of righteousness, tried to destroy the knowledge of the Inca, Aztec and Mayan peoples in South America. The Nazis were guilty of the same crimes, when they burnt books in Germany during the Second World War. The Taliban are reputedly destroying Buddhist statues in Afghanistan today.

This loss of earlier knowledge and skill is evident in historical times that are closer to our present day than the Ancient Egyptians. Take the dark ages in Europe for example. After the Romans left Britain and their Empire was collapsing the technical and organisational civilisation that they brought with them declined rapidly and the remarkable knowledge that they had employed and had maintained for hundreds of years disappeared as the country descended into chaos. A simple example of our British ancestors' declining ability and skill is under floor heating. The Romans used to install it as a matter of course, but upon their demise, it was to be lost for over a thousand years.

What does history tell us? Natural catastrophe, wars and invasions, religion and new ideas all serve to raise and destroy knowledge and the organisation that goes with it. It proves to be governed by a cyclical effect that is constantly repeated over the ages. Knowledge is built up and then destroyed. The stage is the same, some would say that even the players are the same, it is only the conditions and times that is different. So cycles rule again, the cycles of the life of an individual, the cycles of societies and that of the ever-changing world. The Ancients were afraid of even more dangerous cycles in the great clock of time and set out to measure and predict them. Chaos was the enemy and Order was the friend. Let us briefly discuss the concept of ultimate chaos, not just for the ancients but for us as well in these modern times. Let us visit the greatest story of destruction ever told. The Great Flood.

CYCLICAL DISASTERS

Our geological history tells us that massive catastrophes regularly occur. The current belief is that an asteroid strike was the root cause of the destruction of the dinosaurs and some feel that this may have happened more recently in our history than previously thought. Volcanoes certainly have an effect on both a localised and global scale. Pompeii is an example of this destruction, and there is thought amongst some scholars that the ancient Egyptians were affected by the eruption of Thera in the Mediterranean around the time of Akhenaten.[xii] It has recently been discovered that the eruption of Thera was much larger than previously estimated. So large that it affected the weather of the planet by throwing great quantities of ash and sulphuric acid into the upper atmosphere. This terrible event caused what we might think of today as a nuclear winter and it is suggested that starvation resulted on a vast scale over a wide area of the Northern Hemisphere due to crop failures. The evidence of this, the second largest volcanic explosion in the last 10,000 years is found in tree rings in the bogs of Ireland and in ice core samples from Greenland. The ancient oak tree rings show several years of stunted growth, indicating either excessive cold or extreme rainfall or both. The Greenland ice cores show considerable quantities of sulphuric acid as a fall out from the Thera explosion. It has been postulated that this eruption destroyed the Minoan civilisation that had a large marine trading area around the Aegean Sea, which was centred on the volcanic island of Thera. Although there is evidence that the Minoan physically survived the disaster, as later clay tablets and potteries inscribed with depictions of sea creatures indicate it is suggested that they lost faith in their priests and religious systems. Their priests, who lived in vast palaces, were said to have control over nature, much as we think that we have today. The society of the Minoan fell apart as the old order was replaced with a new way of thinking that centred on the mighty powers of Nature. The Minoan started to sacrifice their children as an appeasement to the Gods of Nature. I believe that the same event affected the Ancient Egyptians and this was the cause of the breakdown of the old religious order in that area of the world at the same time. Akhenaten was the heretic Pharaoh, who managed to persuade the population of Egypt to worship the Aten as superior to Amun. I believe he did this because he was suddenly brought face to face with the reality of the creative power of the sun. If the sun was darkened and the explosion of Thera affected the crops, then it would become obvious where the source of life was based. Many friezes in the time of Akhenaten show the Aten (Sun) with rays extending down to Earth with hands holding the symbol of life, the Ankh.

The meaning of the name Akhenaten is the *"Enlightened one of the Sun"*. Akhenaten tried to show this event in self-portraits by depicting himself; his Queen called Tye and their family, as having thin bodies and swollen stomachs. Many scholars express surprise at this, as previous Pharaohs show themselves in an idealised way, handsome, broad of shoulder and athletic. Some try to say that Akhenaten was diseased or trying to show the female side of his nature. However, I believe he was showing how the lack of sunlight caused by disaster, had caused starvation amongst his people and that the effects could be displayed by the swollen bellies of malnutrition. This is an image we all too frequently see today in the television images of the war torn and starving populations of some third world countries. There is also the belief that this volcanic event was recorded in the story of the plagues of Moses, prior to the Exodus recorded in the Bible (See *Act of God* by Graham Phillips). Akhenaten's reign did not last long, he disappeared and his city, Akhetaten meaning *"The Horizon of the Sun"* was eventually abandoned as the old priests of Amen regained power. This disaster may not be seen as cyclical however there is the present danger of the super volcano under the Yellowstone park region in America erupting with even more devastating results.

The Flood is also written of in the Bible[xiii], The Koran, The Dead Sea Scrolls and the Torah, as well as being documented and talked about in international ancient history and myth all over the world. There is little doubt now or conflict between science and ancient writing that the sea levels rose suddenly and dramatically during Neolithic times. This was a terrible catastrophe for humanity, as human beings tend to live and build cities close to the sea as most countries, even in our modern times, depend on sea going trade. No wonder evidence of past civilisations are hard to find, they are mostly under hundreds of meters of seawater. Should such a likely event and I mean likely, happen in our modern western civilisation with its delicate infrastructure, an enormous proportion of the population of the civilised world would be displaced at the very least. Where would they go?

Modern nuclear power stations are built at sea level for cooling purposes. According to a recent report released by British Nuclear Fuels, it takes ten years to shut one down and one hundred to make it safe. How will they protect these highly volatile nuclear furnaces from sudden sea level rises? Our present civilisation is becoming more and more likely to be lost forever through the effects of global warming upon the planet on which we live. This global warming is already seeing a rise in temperatures that may melt the land-based ice in Antarctica. Should the world temperatures rise by an average level of 5°C, then the sea levels will rise by 25 meters. Most of us either fail to remember or have little or no understanding of the methods of travel our ancestors had available to them. The majority of us inhabit countries today that have the benefit of substantial networks of roads, air travel and railways, and we often forget that our ancestors had very few, if any, of these modern amenities. The most obvious and preferred method of travel in ancient times was by sea.

Traders would have plied their wares between nations using this method of transportation; so consequently, most cities were sited at river estuaries, just as they are today. The geographical position of our modern cities is a legacy we have retained from our sea going past. This logical siting applies all over the world: London, Shanghai, Amsterdam and New York are all just above sea level at the mouths of some of the world's mightiest rivers. However, should such extreme events happen today and the land-based ice was to quickly melt in the frozen Polar Regions, the results would be devastating. All these major cities would be completely destroyed by the waves and several thousand years from now, it would be difficult for future historians/archaeologists to find any traces of our civilisation, as most of it would have rotted away under the water.

Scientists are now telling us that vast sections of Polar ice did melt, raising sea levels and thus ending the ice age in only relatively recent times. They now believe that it was so recent as to have been only 10,000 to 12,000 years ago. It was at this time that a large percentage of the life forms existing on the Earth became extinct. There was the total annihilation and extinction of huge numbers of giant animals such as the Mastodon, Sabre-toothed Tiger and Giant Elk. These marvellous mammals walk the Earth no more. Mankind at that time had cause to hunt these creatures for survival and make clothing from their skins. Is it possible that we are in fact, the descendants of the survivors of a terrible catastrophe that had wiped out most of the life forms on our planet as the ice age ended? Is there a group of people among us who still carry an ancient and inherited knowledge that was religiously passed on from generation to generation? Are these select people aware of a cyclical, predictable system of events that we are not privy to?

The secrets that I am about to reveal to you in this book may also be the occult knowledge of a few present day Gnostics, having been kept hidden from the masses and passed down the ages through generations of initiates and secret societies for thousands of years. Of course, it is also reasonable to say that these secrets may have been lost or forgotten through the passage of time, social change and conflict. Perhaps only a few fragments of this knowledge are still hidden away in the cobwebbed vaults of dusty, religious libraries around the world. It is possible that they are completely understood at the highest levels of the powerful present day secret societies and the Roman Catholic Church. These secret societies include Knights Templar, the Freemasons, the Rosicrucians and the Priory de Sion. I will uncover some more information about some of these societies later on in this work.

There is no doubt in my mind that we have lost a relationship with the Earth and the cosmos that our forefathers once understood with all their faculties and wisdom. A wonderful, personal, intimate but tenuous relationship that is still partially retained and honoured by the indigenous peoples of the world. I intend to prove that Mankind was once capable of travelling the oceans of the world, thousands of years ago, before recorded history and that they traded without restriction, as free as the wild birds still do today. I will further argue that they developed skills and instruments capable of telling the time and finding positions on the Earth and sea. That the most powerful of these instruments was the Celtic cross. I also will continue to show that the ancient peoples understood and lived in harmony with the cyclical changes that affect our planet. Changes that happen not just seasonally, but over millennia. That they knew of cyclical, violent and sudden changes that could be disastrous for mankind. These changes were astronomically and astrologically predictable and explain the prophetic abilities of lost races like the Egyptians, Sumerians, Babylonians, Maya, Aztec and Inca.

Why did our forefathers build their incredible stone structures, for what purpose? Could our ancestors have predicted current global warming and its effects on the planet? As I have already said, the Mayan calendar appears to be accurate so far, with the threat of the Northern Hemisphere being destroyed by fire in 2012. All over the Northern Hemisphere today, flash fires are breaking out with ever increasing ferocity, being harder and harder to put out. Were the people who devised this calendar aware of the 11-year sunspot cycles? Did these ancient peoples understand the power of the Sun? Lets see what the scientists have to say about the latest discoveries on the nature of the sun in our times.

THE SUN DOES A FLIP

NASA scientists who monitor the Sun say that our star's awesome magnetic field is flipping - a sure sign that solar maximum is here.

"February 15, 2001 — You can't tell by looking, but scientists say the Sun has just undergone an important change. Our star's magnetic field has flipped.

The Sun's magnetic North Pole, which was in the Northern Hemisphere just a few months ago, now points south. It's a topsy-turvy situation, but not an unexpected one.

"This always happens around the time of solar maximum," says David Hathaway, a solar physicist at the Marshall Space Flight Centre. "The magnetic poles exchange places at the peak of the sunspot cycle. In fact, it's a good indication that Solar Max is really here."

The Sun's magnetic poles will remain as they are now, with the north magnetic pole pointing through the Sun's southern hemisphere, until the year 2012 when they will reverse again. This transition happens, as far as we know, at the peak of every 11-year sunspot cycle — like clockwork.

Earth's magnetic field also flips but with less regularity. Consecutive reversals are spaced 5 thousand years to 50 million years apart. The last reversal happened 740,000 years ago. Some researchers think our planet is overdue for another one, but nobody knows exactly when the next reversal might occur.

Although solar and terrestrial magnetic fields behave differently, they do have something in common: their shape. During solar minimum the Sun's field, like Earth's, resembles that of an iron bar magnet, with great closed loops near the equator and open field lines near the poles. Scientists call such a field a "dipole." The Sun's dipole field is about as strong as a refrigerator magnet, or 50 gauss (a unit of magnetic intensity). Earth's magnetic field is 100 times weaker.

When solar maximum arrives and sunspots pepper the face of the Sun, our star's magnetic field begins to change. Sunspots are places where intense magnetic loops — hundreds of times stronger than the ambient dipole field — poke through the photosphere.

The ongoing changes are not confined to the space immediately around our star, Hathaway added. The Sun's magnetic field envelops the entire solar system in a bubble that scientists call the "heliosphere." The heliosphere extends 50 to 100 astronomical units (AU) beyond the orbit of Pluto. Inside it is the solar system — outside is interstellar space.

"Changes in the Sun's magnetic field are carried outward through the heliosphere by the solar wind," explains Steve Suess, Because the Sun rotates (once every 27 days) solar magnetic fields corkscrew outwards in the shape of an Archimedian spiral."

THE SERPENT AND THE EGG

Far above the poles the magnetic fields twist around like a child's slinky toy acting just like a serpent in an egg. As you have understood from the NASA article above, the solar system is like an egg with the sun at the centre and it is contained in a magnetic shell that extends outward into space. We know this because Voyager has observed solar flares on its journey out through the solar system. These flares have been observed escaping the sun and travelling out past the Voyager spacecraft and then bouncing back towards the sun again. It seems that nothing can escape this envelope that is made of magnetic energy. The solar system is a closed unit on the outer edge of the Galaxy in which it exists. Directly above the sun at ninety degrees to the ecliptic plane is a serpentine coiling of magnetic fluxes and above that at the same angle is the serpent shaped constellation of Draconis "The Serpent" or Dragon. This is the part of the solar system known as the ecliptic or true pole that our ancestors were aware of and it would appear to be the only way in or out for an entity or soul made from electromagnetic energy. The sun is the most powerful object that is near us and affects the environment and us as it changes its output of incredible energies. The Egyptians and the Mayan people understood what they named as *"Grandfather Sun"*. I do not believe that they worshipped it then, any more than they do today. But they thanked it for the sustenance of life that it gives, in the same way that they thank trees for the shade they give. The Mayan people were able to predict sunspot activity and its cycles and Akhenaten was descriptively aware of its destructive power. We are not and have lost our way with our dualistic and separate divided way of thinking.

Why is it so hard for the experts and governments of the world to see that it is the sun that causes global warming? Our scientists are in disarray over this emotive subject. They are in conflict with each other, one side put the blame for this destruction firmly at mankind's doorstep whilst the other camp believes just as strongly that it not the fault of man. Whereas, it is true that we should treat the gardens of the world with the greatest respect, is it true that we create as much CO_2 emission from modern day living as Mount St Helen's did when it erupted? I don't think so and I feel that no reasonable and logical way of thinking could create such an arrogant view. Do the American authorities know this? Is that why they do not sign up for international agreements on the reduction of so called greenhouse gases? Are some of the Governments in the western world aware of the real reason behind global warming and taking advantage of the ignorance of people to suppress competition from the growing economies of third world countries by creating expensive and sometimes uneconomical greenhouse gas emission controls?

What great knowledge have we forgotten during the centuries of greed, war and religious conflict? Where are we heading with our overpopulated world and the greater demand for material wealth and energy? What long-term effect will this have on our civilisation? I believe that our present troubles started with the development of the agricultural society and the abandonment of the hunter-gatherer traditions of the nomadic lifestyle.

The development of an agriculturally based society

Many years ago, I was taking a ferry to Holland from England. It was late autumn and the cold winds were blowing over a grey, white capped sea as the ferry ploughed across the North Sea to the port of the Hook of Holland. About halfway across when we were in the middle of nowhere, I noticed birds flying alongside the ship. There were all sorts of birds, some gliding, some bobbing up and down, all keeping their speed of flight in time with the boat's speed through the water. These were just some of the many species of birds on their winter migration to the warmth and sunshine of the continent of Africa. These tiny bundles of feathers were making an incredible journey; they had thousands of miles to go to their winter feeding grounds. Some of these tiny-feathered creatures were fledglings that had been hatched this year and others were seasoned travellers. Occasionally, one would settle on the decks of the boat for a rest. I noticed that the majority of my fellow travellers were oblivious to these wondrous creatures as they rested, weary from their long flight of faith and freedom. We were still only halfway across and it was getting dark. There were still miles of unchanging sea and sky ahead of our ship and the birds. There was no sight of land to give hope and comfort to them. I began to think of the problems that they would encounter on landing in the dark, on an unknown shore, many for the first time. After all, most of these species roosted at night and were not known to fly around in the dark. But they could navigate, no one knows how, but even the young would reach Africa and those who survived the journey would return next year to breed a new generation. Repeating the cycle over and over again.

It was only on arriving in Holland that I had a revelation. I had passed through customs with my passport in hand, reading notices ordering us not to import or export livestock and seeds. But here they were. the tiniest of God's creatures, with their tiny bodies and frail bones, doing just that. They had no passports, they did not have to go through immigration and the 'Anything to Declare' channels. Here they were, importing themselves and the seeds that they had consumed in England straight into the soil of Holland or Belgium without a "by your leave" from the authorities. They were truly free and unrestricted. Whereas, I was constrained by language and nationality, carrying my papers just so I could cross borders that were named and designated. I realised at that moment that I was living in one world, The World of Man, whilst the majority of the other creatures inhabiting the Earth were living in another, The World of Nature.

This other world would never come to an end, but ours was so tenuous and fragile that it could quite easily. Consider the way in which the majority of the population of the western world live. They live in overcrowded cities and modern housing estates; these housing estates have no facility or garden space to grow food. Food comes from supermarkets supplied on a "*just in time*" management basis and most of it is imported from other countries. Most modern supermarkets have enough stock for only a few days. The supplies are so tenuous, that if there was a break in the supply chain for any reason, the people would starve and anarchy would break out very quickly. If there was a break in supply, people can not purchase food from the local farming communities any more as economics have forced the arable farmers into growing inedible cash crops such as sunflowers or oil seed rape. Worse still, there are major corporations involved in genetic engineering, which is deliberately building in a genetic inability in grain. This inability is designed so that the plant will not produce grain that is capable of re generating itself the following year. The concept behind this is to force arable farmers to become less self-sufficient and purchase all their seeds from the corporation. What if, for any reason, the supply chain breaks down? This is material madness beyond belief.

How had this state of affairs come about? It is all the result of successful agricultural methods. As soon as a society settles down and starts to farm the land there becomes a requirement for measurement. This measurement was used in Egypt so as to re-measure the land to re establish ownership after the fields were washed away by the annual Nile flood. Measurements in Egypt were generally known as cubits. A cubit was considered to be the length of a forearm. There were different levels in sophistication of measurement. There were those designed for the farmer, who was considered to be at the bottom of the social hierarchy and another for priests, architects, astronomers and Pharaohs. The Royal Cubit was sacred and was applied to building. In order to be able to tax the farmers, one must be able to measure their wealth. Scales with a system of weights and measure were employed. To layout the fields, to plant and reap, the seasons must be defined and predicted. Techniques in measurement were applied to this and geometry was formed. The cross and plumbline was used for more sophisticated types of measurement, such as astronomy for time keeping and surveying for building. Success in agriculture produces wealth, sometimes to excess. The consequence of this excess wealth, as agriculture grew, was an ever-increasing population and expansionism.

Initially, there was a continuation of free trade but this deteriorated as armies were used for attack and plunder rather than defence. Neighbouring communities started erecting borders and armies were established for security purposes.

A measure of border control was implemented and restriction in travel became the norm. As communities fell and were absorbed into the growing territory of the aggressors, countries were formed. Most of these new countries were restricted to natural boundaries, such as rivers, coastlines and mountain ranges. The population was encouraged to become patriotic and fiercely territorial. This is where we stand today, with a growing population and dwindling resources. This state of affairs is untenable in the long term and our modern day governments try to find ways to solve this problem financially and in control of consumerism. Many may feel uneasy as changes hurtle them to God knows where. To know where we are going, first we must know where we came from.

I want you to imagine a very different world thousands of years ago. A world where humans were known as "The Children of the Earth" A traditional hunter gatherer society. These people were free to roam the Earth and followed migrating animals as the seasons changed. There are some indigenous people living in remote parts of the world who still live like this. I also believe that there were those who lived this lifestyle on the oceans of the world before and after the Great Flood and that all the knowledge of the Ancients came from these mariners who survived the chaos and destruction of the melting of the ice age. I believe that these ancient navigators knew astronomy so that they could navigate the world and that they had an instrument for measuring the stars and telling the time. I am convinced that the instrument was the cross with a plumbline.

PREHISTORIC OCEAN TRAVELLERS

Our ancestors were Sea People and travelled the oceans with the freedom of the Albatross using the wind for power.

"Wind"

"What sees the wind?"
In soaring climb to heights that touch the very face of "God."
"Life", it whispers to the newly tender, freshly born in softened, warming spring.

In gentle play, caressing roses in my summer wilting garden, with scented promises of rain.
"Death", screams and shrieks a violent dance in shattered, leaf crushed forest as groaning, ancient giants fall.

Roaring tumult of the Deep breaks ships of men and shells of fish on distant, daggered shore.
Still, still, grips the silent crystal night in deathly, frozen, waiting sleep.
Ghostly, are the hands of unseen air that move the face of ancient stars that watch the cycles of the World.

Far, far away and close.
The breath of distant ancestors and every living thing that ever was and is,
are carried on the wind.
It is I and you and what will be.
Forever
Eternal is its name.

© 2000 Crichton E. M. Miller

The wind is the oldest and most powerful means of harnessing power and our ancestors harvested it thousands of years ago to drive their ships of trade and war. They built ships of grace and beauty, craft that were at one with nature and used the elements in harmonious style. It is well known that they crossed seas such as the Red Sea and the Mediterranean over four thousand years ago and archaeologists find new evidence of this skill dating further back than previously reported on a regular basis. Let us deal first with the possibilities of ancient transoceanic travel. Is there any evidence of intercontinental pre-Columbian Voyages? Learned authorities dictate that no one was able to cross the Atlantic Ocean before Christopher Columbus discovered America. It has to be said that this is the opinion of the general population in light of the available evidence. This public view is also maintained in academic circles despite the efforts of such stalwarts as Thor Heyerdahl. Thor Heyerdahl is the famous Norwegian ethnologist who set sail on the ancient Peruvian style raft, Kon Tiki, from Peru to the Tuamotu Archipelago and also embarked on a transatlantic crossing in an ancient Egyptian Papyrus style boat in the Ra Expedition.

Many people still like to cling to the old belief that there was no cultural interchange between intercontinental cultures. Having dedicated his life proving that the ancients had seaworthy vessels capable of these long voyages, the current argument against Thor Heyerdahl is based on the idea that the ancient mariners were incapable of navigation. They claim that migrations of the ancients between the continents were via land bridges, and over vast periods of time. If this line of thinking is correct, then certain questions must be asked. Why are there Step Pyramids in Egypt, Sumeria, Tenerife, Mexico, The Yucatan, Peru, Chile, China and recent ones discovered in Mongolia? The answer, according to some academics is simple; it is all just a matter of plain coincidence. Similar cultures sprang up in disconnected parts of the world at different times by sheer luck rather than design. After all, diffusionism is a dirty word today.

We are expected to believe that the various cultures around the world, all of which demonstrated similarities, evolved independently of each other with no interchange of knowledge. I can assure you, that if such coincidental evidence were offered in a court of law as a defence, the accused would be convicted every time. This is a preposterous representation of the evidence. It is based on the idea that most of the artefacts bearing a resemblance to each other were tombs or sacrificial sites and not the highly sophisticated instruments which I am about to reveal to you.

A prime example of these constructions would be the incredible Pyramid of Chichen Itza in the Yucatan, with its Astronomical Light and Shadow Phenomena. If the existence of these incredible structures is just plain coincidental, then how do you explain their similarities when compared to the Giza complex in Egypt?[xiv] It would appear that even though they are miles apart, they all seem to have the same methods of construction, the same beliefs and a vast amount of skill in astronomy, which assisted them with their buildings. We are supposed to believe that these structures and civilisations were independent of each other yet we are unable to replicate such a construction project today, let alone use one.

There are even more coincidences that should be looked at. There are Mummies in Egypt, Tenerife, China and South America. How do you explain the evidence of ancestor worship in Egypt, ancestor worship in the Canaries and ancestor worship in South America? There is also evidence of a high level of knowledge and understanding of astronomy and astrology in Egypt, evidence of astronomy in Europe and again, evidence of high astronomy and astrology in South America. Surely this has to be more than mere coincidence.

Furthermore, there is evidence of the art of Prophesy being utilised in Egypt and the Middle East, as well as in South America.

SIMULACRA AND ANIMISM

Likewise, Simulacra was used in Egypt as it was in South America. Following suit, you find that animism was used as spiritual understanding in Egypt and used as spiritual understanding by indigenous peoples of the Americas. Maureen Palmer is a researcher and friend who has studied the ancient beliefs of Simulacra and Animism. I shall let her explain the concept with her remarkable discovery in both Egypt and England of the images of *"Sacred stones"*.

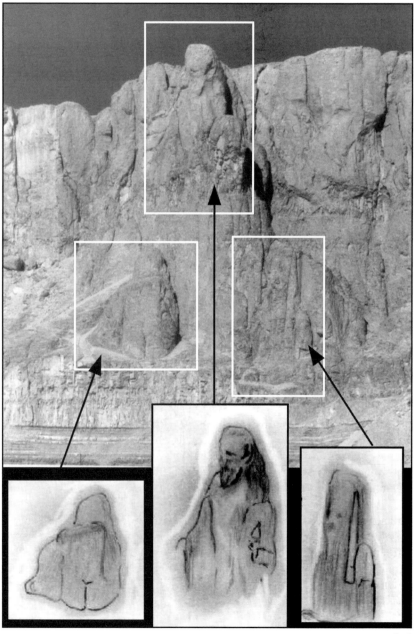

fig.1

Sacred Stones

By
© 2001 Maureen Palmer[xv]

There is a phenomenon that was recognised and widely venerated and is still recognised by many older native cultures today but is little studied or valued within our more modern scientific worldview. This phenomenon is called simulacrum (plural). A simulacra is an image perceptible to the human mind and seen as a life form resulting from a combined mental process whereby our brains process patterns of shadow and light reflected from physical objects. Nature has created a plethora of these images especially in rock formations, frequently found in very close proximity to prehistoric monuments, temples and ancient burial sites around the World.

Animism, in ancient times, was common to almost all indigenous people, globally. Animism is a belief that an animating force fills our Universe with life and spirit. Sir Edward Burnett Tylor, an English Anthropologist, first used the term Animism. Born on 2 October 1832 in London England, Tylor became head of the University Museum at Oxford. He was Anthropology Professor there from 1896 until 1909. His work made great advances to our understanding of indigenous religions. Animism regards every plant, tree, mountain, stone, living creature, sky, and ocean, celestial body etc. as having a soul, and any animate or inanimate object can serve as a sanctuary for a spirit, which can be either good or bad. The Shaman attempts to communicate with these spirits in order to persuade them to cooperate. Encounters with spirits and natural phenomena survive today in mythology, as angels, demons, fairies, sprites, gods and supernatural beings.

In ancient times the Universe, the Earth and the Environment was considered worthy of reverent wonder and worship. Historically, through all known history of humanity, the Sun and Moon were revered as the most important primary requirements of all forms of life on Earth. The Sun being the body from which the Earth was born, giving us energy and light, and the Moon, because of its mass, a major influence and regulator upon both waters of the Earth and waters within our bodies. Knowledge of celestial movements which in turn heralded seasonal conditions, changing animal behaviours, and the coming of the rains, was critical to survival and the better a society could predict natural events the more successful that society would be.

When the Moon is full, dark shadows can be seen over the plains and ridges that stand out from its general illuminated surface. Moon shadow simulacra have been acknowledged from pre-historic times. They include distinguished human heads, a seated woman holding a child in her arms, a lobster, a donkey and a large rabbit or hare. In Celtic areas of Britain, Germany and Austria, a hare was believed to be a sacred animal of Easter, goddess of the Moon, whose festival was held at the Spring Equinox. In a tradition established in Neolithic times, hares were sacrificed, then axes were offered to Andraste, a war-goddess. When Christianity arrived in the Celtic world Eastre was replaced with Easter com-memorating the resurrection of Christ. The festival is held on the first day of the Sun following the first full Moon of the Spring Equinox.

Stonehenge in Wiltshire England is the World's most famous stone circle and the most outstanding prehistoric monument in the British Isles. At around c.3200 BC Stonehenge started as a circular ditch with a simple central wooden shelter. The next building stages included the building of two incomplete circles. The smaller Bluestones, which were either glaciated or transported to the site, came from the Preseli Mountains in SW Wales 385 kilometres away. The larger Sarsen stones were hauled from the Marlborough downs some 30 kilometres away. The positioning of the stones concentrated on points on the horizon and aligned to predicable cycles of the Sun and Moon. Human bones were found in pits orien-tated to the four main compass points. The north-east entrance it is believed was set to look towards the Horizon where the Midwinter Moon rose at the Northernmost point in a 18.6 year Saronic cycle, an astronomical cycle of 6585 days and 8 hours. One of the Trilitheon Sarsens has numerous axe-heads carved into it; another has what is said to be a Mother goddess carved in relief. Many other stones bear images of distinguished human heads, some of which are only visible at certain times of day or year when the angle of light is just right. Some of these heads are natural simulacra, some are carved.

Known under various names, the Mother goddess is often referred to as the Mother of the Tribe, Mother Earth or the Mother of all that has been, that is, and will be, she was neither male nor female but both. One common image of this androgyny in art form is a bilateral figure with the Male half-facing left, the Female facing right.

At Stonehenge, in the centre of the Inner Sarsen horseshoe, built into the struc-ture and aligned to the North and South compass points is a bilateral androgyny figure with the Male half facing left and the Female facing right. This image has gone unnoticed for generations but it is there for all to see.

On the other side of the Atlantic, popular amongst the peoples of North America were totems, which symbolised a particular family, tribe or clan. Totems can represent an object, a phenomenon of nature, a species of animal, a plant or a bird. The Eagle is a totem that has inspired American poets, mystics, leaders and warriors for generations. The Eagle represents a free spirit and is a symbol of power.

In an area of the Missouri River basin, is a hunting ground used for millennia by native peoples for camping and hunting in seasonal patterns. Recently an American Geologist found two remarkable little stones in this area. The first has simulacra of a soaring Eagle within a stone and on the reverse side, two more wings are visible with other markings below, resembling a man throwing a stick into the air. The second stone would appear to be a worry stone discovered close by. Both stones are very hard orthoclase plagioclase with quartz with the worry stone having been down into the crust several kilometres where it was re-mineralised without losing its banding and then geologically uplifted to the surface. One can only imagine of what value these two stones might have had to the native peoples.

Arizona contains some of the most beautiful natural geological landscape features and simulacra in the world. In NE Arizona, the Hopi (meaning peaceful) people still hold sacred the ground of the ancient ones, as their remote high desert location and isolation has allowed them to retain some of their own cultural heritage and traditions much longer than the larger more commonly known tribes.

In his book 'Road in the Sky', the late Dr. George Hunt Williamson explored the remarkable monolithic carvings and simulacra in rock formations in Peru and Arizona. High on the Marchhuasi Plateau overlooking the village of San Pedro is Peca-Gasha an enormous carved head, countless simulacra, and gigantic carvings of human faces representing all the races of mankind.

In the folklore of some of the tribes of North American, the Chenoo were believed to be spirits that were able to camouflage themselves by blending into rock formations. The Chenoo took the form of great stone giants, both animal and human.

On the other side of the World is Egypt, a land of profound natural beauty and incredible contrasts, with crystal clear blue skies and limestone cliffs that have turned golden in the strong Egyptian sunlight, they gleam and shimmer as the layers of hot air build up around them. Flourishing in agriculture, the land is sustained and nourished entirely by the cool waters of river Nile, with vast scorching desert wastelands on both sides and virtually no rainfall, it is truly a land of blue and gold.

The ancient Egyptian priests claimed that Egypt was an image of heaven, a place where all the powers and operations that work and rule in heaven, had transferred to Earth below. They believed that certain animals, creatures, birds and many inanimate objects possessed divine powers and they gave form to their sacred animals and objects by representing them in human form as anthropomorphic beings.

The Great Pyramid of Giza in Northern Egypt was perhaps the greatest of the Seven Wonders of the Ancient World and the only one still standing today. The Giza Pyramids and the nearby Sphinx never fail to leave the visitor both impressed and mystified.

Dr. Farouk El-Baz, Director of the department of Remote Sensing at Boston University in Massachusetts USA has a long and distinguished career in Geology. Dr. El-Baz studied natural landforms and exposures in the shape of inverted boat hulls, with their pointed ends downwind out in the Western Desert of his native Egypt. As the aerodynamically shaped features were subjected to further erosion, the resulting landforms have sphinx-like shapes of reclining animals. Dr. El-Baz felt that the uppermost part of the Sphinx might have started as a wind-carved feature, whose shape was completed by the ancient stonemasons having been inspired by the natural landforms.

In the South was the heart of Egypt, the ancient city of Thebes, which for much of Egypt's long history was the centre of religion and worship. On the West Bank is the Necropolis, city of the dead. The Necropolis is a timeless place, beautiful and enchanting with deep valleys, high mountains and golden cliffs. It is a place that records the spiritual beliefs of a civilisation, a place that intensifies the senses and a place that makes one question one's own beliefs and perceptions. The most prominent geological landscape feature dominating the Western horizon is a mountain that overlooks the Valley of the Kings. The ancient Egyptians regarded this mountain as sacred and referred to it in the texts as 'The Horn of the West', and described it as a 'gate' and a place where souls, gods and spirits dwell, and what they do there. The sacred mountain is a natural geological pyramid and during the New Kingdom, the peak of the mountain was also venerated as a goddess, whose name signified 'the beloved of him who makes silence'.

The Esna creation story tells how Neit, an ancient androgynous Mother Goddess performed the creation of the world by emerging as a mound from the abyss, the primeval waters of the Nun. It was upon this mound that she brought into being all the other gods and created light by giving birth to the Sun god. Today nothing remains of her temple where she was worshipped. However, an inscription in her temple read: "I am all that has been, that is and will be. No mortal has yet been able to lift the veil, which covers me.

Behind the Western cliffs, nestled in the foothills of the Libyan Mountains is the Valley of the Queens. The valley houses more than eighty tombs of Queens and Royal Children. The ancient Egyptians dedicated this valley to the goddess Hathor, who was perhaps the most complete reflection of the female principal. As the patron of women, female animals, nourishment and new life, she was depicted as a divine cow or as a woman with the ears of a cow. In her funerary role, Hathor was given the title the 'Lady of the Western Mountain'. In this role, she is depicted on papyri vignettes as a cow peeping through a papyrus thicket, half-emerged from the Libyan Mountain. Standing before her is the hippopotamus goddess Taweret, a goddess of childbirth, maternity and suckling, whose male consort was the crocodile god Sobek. At the far end of the valley is a small depression. It is known that in ancient times the priests of the Necropolis filled this depression with water and planted papyri there, seemingly a very strange thing for the ancient priests to have done. However, the reason why becomes crystal clear if one observes closely the rocks directly behind the depression. In the natural formation of the rocks are simulacra of three animals, a cow, a hippopotamus and a crocodile.

Hidden deep in a remote valley close to the village of Deir el-Medina is a rock cut shrine dedicated to the god Ptah. The shrine consists of inscriptions cut into an outcrop of rock. There are no other monuments or inscriptions at the site and perhaps for this reason the site is seldom visited. Ptah's female consort Sekhmet whose name means 'the Powerful' was depicted as either as Lioness or as a woman with the head of a Lioness. In their temple, they were represented as a triad with their Lion-headed son Nefertum. In the natural formation of the rocks directly above the shrine can be seen a simulacra of a 'double-Lion'. The image is that of a male and female Lion sitting together one behind the other in profile, with the Lioness seated in front and her paws stretched out in front of her, both are in same body, forming a double-Lion.

At the northern end of the Necropolis is a place which from very ancient times was given the name Djeser meaning 'the Holy' and it is here that Hatshepsut a remarkable eighteenth dynasty Queen who ruled Egypt as Pharaoh in her own right, built her mortuary temple. To say this site is breathtaking is an under-statement. In the background is the sacred Pyramid Mountain that overlooks the Valley of the Kings. When Hatshepsut built her temple, she dedicated it to the god Amun the chief god of Thebes and renamed the site Djeser Djesru, the Holy of Holies. Amun, his female consort Mut and their son Khonsu were usually rep-resented as a triad in human form.

In the rocks directly above the temple are simulacra of three people (see fig. 1). To the left is an image of a woman in profile, she is seated with her knees raised and her arms wrapped around her legs and her head is resting on her knees, she appears naked with her bare bottom facing forward. To her right is a simulacrum of a young male, the side of his head appears to have the long heavy tress of the typical Egyptian royal child. Above the two, in the centre and reaching the very top of the cliff is a simulacrum of a dignified elderly bearded man. He appears standing and dressed in a long flowing robe.

Today, simulacra attract public attention with almost uncanny synchronicity. This was demonstrated in 1997 whilst the world was still mourning the death of Diana, Princess of Wales. Shortly after her death a silhouette of incredible likeness of her was observed and photographed in damp plaster on someone's patio wall in Australia. Shortly after this event yet another simulacra of her face was seen and photographed on the other side of the world in England, this time ironically amongst the leaves of a tree growing on the tiny island on her family estate.

This same uncanny synchronicity occurred just as science started to seriously look for the existence of intelligent life elsewhere in the Universe and at the same time as we started to send spacecraft out to investigate and photograph nearby planets. Photographs taken by the Mars Viking 1 Orbiter on 25 July 1976 from a distance of 1162 miles showed to the world a human-like face formed in a rocky outcrop measuring one mile across. The photograph of the 'Face on Mars' appeared on television and in magazines and newspapers across the world, creating an almost chain-reaction of speculation.

Gemstones, crystals, worry stones, some sedimentary rocks known as oracle stones, and astral baetyls have been regarded by many cultures for millennia as having spiritual or magical significance. Even today, many are still admired and even regarded as precious. Beaded gemstones threaded together to form a string of beads had a similar purpose to ancient peoples as modern-day rosary beads used for prayer by the people of the Roman Catholic faith. It is said that to rub a worry stone or pass the beads through ones fingers will ease away nervous tension or aid memory. Worry beads are still quite common in Greece and parts of the Middle East.

In future times, Humanity will wander amongst the stars and galaxies. Already we see there are masses of interstellar dust and gas, visible as areas of darkness and reflected radiation and known as Nebula. With the aid of modern technology, it is now possible to see simulacra which are created inside Nebula, some are the most incredible, beautiful and colourful ever produced in the Universe.

If you would like to know more about Maureen Palmer's research work, or see simulacra that she has photographed please visit her website.
The website address is http://www.godsintherocks.com

Co-incidences ?

So let us continue on the road of investigation and I will present some general evidence against the argument that this is coincidence. The question was: "Were the ancients capable of sailing and navigating the oceans in prehistoric times?" We know from archaeological evidence that they were capable of building and sailing ships that utilised wind power. We will find out that they were capable of finding their position on the Earth in terms of latitude and longitude. But first I will present the coincidental evidence of cultural and spiritual contact between civilisations on both sides of the Atlantic Ocean. Some of these civilisations were separated in time by thousands of years but they were all in existence before Columbus discovered America. So, I will present this evidence of contact and similarity on both continents as though it were a court of law.

As Maureen Palmer has ably shown, there is evidence of simulacra in Ancient Egypt and Britain. In the sacred valley of the Inca in South America are simulacra. Their most sacred signs were devised as simulacra of the dark shapes seen in the Milky Way between the billions of stars. The Inca people used the Milky Way to prophesy the coming of the Conquistador and they allowed themselves to be destroyed without fighting back, despite their fearsome military ability and their overwhelming supremacy of number. It was written in the stars and therefore the outcome was inevitable.

It was by these dark shapes that they measured time and prophesied. These three signs were the Toad, the Condor and the Llama. The sacred valley was chosen as a place to settle by their Shaman or wise ones because these images were to be seen in the rocks. There are three mountain peaks and each is similar in shape to each of the sacred creatures. The Inca people went further though; they enhanced the shapes of the Toad, Condor and Llama with terracing. The Inca astronomers could tell the time or the season by the shadows of the sun as it struck these images and they studied the night sky in remarkable depth. They also went on to build an invisible pyramid using their ingenious geometry. It was and still is invisible on the valley floor until it is highlighted by the shadow and light phenomenon created by the sun. They used an instrument to measure the angles; this instrument of knowledge was what they considered to be a staff of power, the Cross. But where did it come from? It must be remembered that the Inca civilisation was separated in time from the Egyptian Old Kingdom by thousands of years. Furthermore, there were thousands of years separating the Egyptians from the Neolithic megalith builders. Thoth is indeed reincarnated here and there throughout the history of Mankind as is told in the Emerald tablets.

Henges and megaliths are found throughout the world. The most famous henge is Stonehenge in Wiltshire, England. There are henges in wood and stone to be found in Scotland and England. Henges and megaliths are generally found at higher latitudes than pyramids. Megaliths are found in higher latitudes both north and south of the equator. There is a megalithic gate of incredible antiquity in Tonga, South Pacific Ocean. There are megaliths in the Marquesa Islands, found by Thor Heyderhal. In fact, the discovery of these megaliths was part of the inspiration for him to embark on his remarkable career. I have discovered that these pyramids and megaliths are the clocks and calendars of the Ancients.

What is the significance of the Serpent?

Images of serpents have been found in Egypt, Scotland and in the North and South Americas. The drawing above is based on a sketch made during the last century from the banks of Loch Nell, near Oban in the West of Scotland. It depicts a manmade form in the shape of a serpent. Below is a photograph of how it looks today.

This earthwork is on an enormous scale and must have required incredible effort on the builders' part. It is aligned NE by SW. It is relatively unknown to the locals although they have heard of its existence. The local people that I spoke to, who lived near it, had thought that the serpent mound was small and did not realise the size of it until I pointed it out. The curators at Kilmartin Museum had also heard of it, but they had not yet got around to investigating it. However, it is not unique, as a serpent mound can also be found on the eastern seaboard of America, in Ohio. So again we have culturally similar prehistoric manmade constructions on both sides of the Atlantic. Dragons and serpents in South America and China are seen as ancient symbols, depicting the Earth Spirit.

The Bible in the Book of Genesis Chapter 3 mentions serpents in the narration of Adam and Eve in the Garden of Eden. Eve was encouraged to eat fruit from the tree of knowledge of good and evil after having been instructed by God that this should not be done for they would have to suffer the consequences. The serpent persuaded Eve that this was not the case, and having seen how tempting the fruit appeared, she ate the fruit and then persuaded Adam to do likewise. God, true to his word, punished them both for having disobeyed his commands and to the serpent he condemned a life of living on its belly (chapter 3, verse 14) and declared it to be '*cursed above all cattle and above every beast of the field*'.

The imagery of serpents is to be found all over the world. Depictions of serpents can be found on the headdress known as the "*Nemes*" worn by the Pharaohs. Such an example is the nemes of Tutankhamen, which shows a cobra beside a vulture. Serpents play a large part in Amerindian folklore. The Maya people talk of the Rainbow Feathered Serpent and attribute great powers to this ancient deity. In Britain, Middle Ages cathedrals such as Lincoln show carvings of serpents with the vine of life issuing from their mouths. The serpent or dragon is also shown in Roslyn Chapel and particularly on the base of the Apprentice Pillar[xvi]. The Chinese people display dragons and serpents in their literature and in street theatre even in these times.

It therefore has to be asked, what is the significance of the serpent to these ancient civilisations?

The Constellation of Draconis

I believe that the serpent or dragon was one of the most important astrological and astronomical constellations in the sky. It was important because it was part of the circumpolar *"Immortal Stars"*, the stars that never set. It is the constellation of Draconis in the Northern Hemisphere of the sky. I believe it was used for timekeeping, astrology, astronomy and navigation. It is possible to work out where the sun, planets and the constellations of the Zodiac are by observing the position of Draconis in the night sky. Its star patterns act as pointers in the same way as the Plough and Leo. The following plate shows the position of Draconis in relation to the North Pole on the 12th of July 2001.

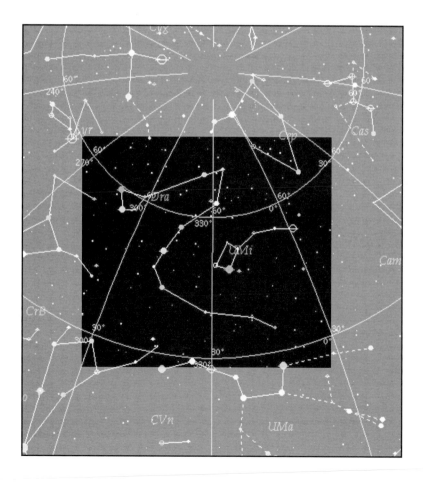

The following is part of a translation from The Egyptian Book of the Dead by Wallace Budge.

The gods are glad [when] they see Ra (The Sun) in his rising; his beams flood the world with light. The majesty of the god, who is to be feared, setteth forth and cometh unto the land of Manu; he maketh bright the earth at his birth each day; he cometh unto the place where he was yesterday. O mayest Thou be at peace with me; may I behold thy beauties; may I advance upon the earth; may I smite the Ass; may I crush the evil one; may I destroy Apep)[1] in his hour; May I see Horus in charge of the rudder, with Thoth.

Apep, the serpent, personifying darkness, which Horus or the rising sun must conquer before he can re-appear in the East. This ancient book describes how the rising sun (Ra) at dawn banishes the constellation of the serpent Draco (Apep) from the night sky. Bringing light and banishing darkness. We have further hidden clues to navigation from the Old World to the new. A mystery that links Africa and America in prehistoric times, is the difficult question as to why traces of cocaine and tobacco were found in Egyptian Mummies by archaeologists when it is a known fact that cocaine originates only in South America. How did it get there?

In finding an explanation to these mysteries, you have to turn to the legends relating to these countries. There are legends of white bearded peoples arriving in ships from the East in the South American Indian myths, an example being the God known as Viracocha[xvii] from the Incas. Legends speak of a bearded white man with blue eyes and fair hair who brought super knowledge to the Amerindian People. He arrived from the East and taught them how to build Pyramids and stone buildings. He taught them the skills of agriculture, astronomy and mathematics. He was known as the rainbow feathered serpent and called Quetzecoatl, Ku-Kul-Kan, Hyustus, and Kon Tiki amongst the Aztec, Maya, Toltec and Polynesians. The Maya still call on Ku-Kul-Kan today in their ceremonies. Quetzecoatl and his people are said to have returned to the east, promising to return one day. This is the legend that destroyed the Inca and Mayan peoples as a civilisation. For their prophesy was fulfilled the day the Conquistador arrived. So here again the serpent shows its important integration into worldwide myth.

Lord Pacal,*xviii* the priest-King of the Maya who lived, died and was buried in the Temple of the Inscriptions at Palenque, was known as the feathered serpent. Here again we have a similarity with Ancient Egypt, which will become clearer as we move on. The Egyptians and the Maya both saw their priest kings as Star Gods manifest on the Earth in human form. These God-men, separated by thousands of years, were Gods of Time, because they knew how to tell the time through astronomy and astrology and they originated in the East across the Atlantic Ocean. Who could these people of great knowledge have been? Travelling up through the ages from the earliest of times the list is quite large, however it is unlikely that the earlier names on the following list are the Quetzecoatl of Mayan Legend because of his blond hair, blue eyes and beard.

With all of these similarities, surely we cannot be wrong in assuming that there had been some form of intercontinental travel, a sharing of beliefs, knowledge and skill between these civilisations separated by time but which never dies forever. The list of potential mariners embarking on trans-oceanic voyages could be comprised as follows: Firstly there are Neolithic mariners after the flood, followed by Celtic and Indian seafarers. Egyptians and Phoenicians were known to venture on long sea voyages. In the Egyptian beliefs and Funerary rights there is the expressed desire by the Pharaoh of heading to the beautiful West on a boat after his death.

There is the possibility of a voyage by Akhenaton and his followers fleeing from Egypt. Then, later, there are the Lost Tribes of Israel, where did they go? Followed by the Essenes fleeing from Rome after burying the Dead Sea Scrolls at Qumran. The translation of the word Essenes means a School of Fishes, and there is considerable evidence that these people were aware that they lived at the start of the age of Pisces. Fish head statues have been found in South America, which is too much of a coincidence with the age of Pisces to be ignored. Then, a thousand years later, there are the Knights Templar. The likeliest candidate for Quetzecoatl is a member of The Knights Templar who may have visited South America with a crew several hundred years before the Conquistadors arrived in South America. The story of King Arthur and the Round Table is a Templar based legend. Did you know that Lanzerote, in the Canary Islands, means Lancelot? What is the meaning of this coincidence?

The greatest and yet most frowned on discussion about civilisations across the Atlantic is Plato's story of Atlantis as documented in his works of Timeus and Criteas. This is as yet unsubstantiated as ever having existed, although Andrew Collins in his latest book, *Gateway to Atlantis*, argues a strong case for Cuba and the Bermuda banks as a possible location for such a place.[xix] All these similarities and circumstantial evidence throw our current understanding of our history into doubt. In my mind, ancient mariners definitely crossed the oceans of the world in pre-Columbian times, there is so much evidence pointing towards this and after you have digested my research, there will be less doubt in your mind as well that this was indeed the case.

So, lets go back to the court of law scenario. I will gather the evidence and present it to you to the best of my ability. I should warn you though, you may not like some of the truths that I expose, but you will be able to exercise your freedom of choice in this journey of discovery. You can not fail to be amazed at some of my revelations and in the end, you the reader, will be the Judge and Jury. As in all cases, the only drawback to proving any theory is the lack of physical proof, but in this case, I have found it. There is much circumstantial evidence of transoceanic voyaging and cross-cultural exchange between peoples presented by authors and researchers over the many centuries since Plato. These concepts are vigorously rejected by academic institutions as being nonsense. Why should this be the case?

It is fair to say that the researcher has to exercise considerable discrimination when studying other writers' theories on the existence of the lost city of Atlantis. Some of these theories have brought the whole discussion into disrepute. I believe, however, that one of the real reasons behind this rejection of the Atlantis theory is the harm done by Hitler, before and during the Second World War. He adopted the swastika as a symbol of his monstrous dictatorship and it is becoming more widely known that he was a firm believer in the lost city of Atlantis theory, spending large amounts of resources on seeking out the lost knowledge of this fabled and ancient civilisation. The swastika is an ancient symbol for measuring the Earth and is also to be seen on the Coat of Arms of the German freemasons as I have discussed earlier in this book. Hitler had managed to persuade the German people that they were an elitist race of people and directly descended from the Atlanteans. Hitler's pure race was to become known as "*Aryan*" as in the ancient name given to Atlanteans. The concept of elitism was the true horror of his party and resulted in the Holocaust and mass murder of millions of innocent human beings. This is one of the reasons that I have written this work, made a website and patented the ancient instrument of the cross. It is to avoid the evil of division through elitism and discrimination while maintaining my personal view that all men were created equal. Icons are very powerful instruments and can be extremely dangerous in the wrong hands. The cross must not be used in the same way as the swastika by unscrupulous forces for their own ends. Nor must it continue to be an icon in the hands of religious leaders who have lost the meaning of spirituality. It is my hope that the cross will become a symbol of unification and healing, mending the evil of division. It is the last piece in the jigsaw, the last real proof that the ancient people of the world were as One, capable of telling the time and navigating the world's oceans. I will now present to you the story of my rediscovery or reincarnation of the cross as a real and practical instrument of the ancient mariners, and proof that they were sufficiently competent navigators, able to cross the oceans of the world. I will resurrect for you the instrument hidden away four thousand years ago by Kings descended from ancient mariners, Kings who believed themselves to be the interface between the people they served and the God of the Age they lived in. The facilitators and communicators with nature and creation in the time known as Aries or Amen.

CHAPTER 4

THE
REINCARNATION
OF THE
WORKING CROSS

I have discovered the physical proof required showing that the ancients were capable of a higher form of mathematics than previously imagined by the public and academia alike. Proof that demonstrates the ability of the ancients to tell the time, employ accurate navigation techniques, survey new buildings prior to construction, plus measuring distance and height accurately. If those abilities were the full extent of the instrument's possibilities, it would be amazing enough. However, that would be underestimating its full capabilities.

This magical instrument that I have discovered, hoary with age, is not solely used for navigational purposes. It encompasses a knowledge of the cosmos, the use and understanding of mathematics, geometry, surveying, astronomy and astrology and it was used long before the ancient Egyptians built their pyramids. The secrets of this device were the foundations of ancient civilisations, long before the written word. It was the staff of magicians. I have searched high and low, in libraries, museums and on the Internet, even with experts and engineers but to no avail, no one in the modern world previously knew of its existence. As a result, I did not fully understand the implications involved with my discovery, but later it was to dawn on me the heavy responsibility of my findings. This discovery changes the very perspectives of history and the Christian religion as we know them. Anyone who sees the instrument in its true light, will never look at world history in the same way again. This discovery does not conflict with the ideas and theories presented over the last few decades by alternative history researchers, rather it gives them a real and physical method for the mathematical and astronomical anomalies about which many of them write. Writers like Thor Heyderhal, Graham Hancock, Robert Bauval, Maurice Cotterell, Colin Wilson, Adrian Gilbert, Lorraine Evans and many others less well known have made remarkable discoveries and pursued the difficult task of publishing their findings against fierce opposition and sometimes derision from the "establishment".

My research on this subject began in 1997 after reading Robert Bauval and Adrian Gilbert's book, *The Orion Mystery*.[xx] I was curious about the system employed by the Egyptians to undertake surveying and astronomical calculations, whilst building the pyramids with such precise accuracy. I was convinced that the builders must have employed some sort of simple theodolite and sextant (I use the words 'theodolite' and 'sextant' loosely to describe an ancient instrument capable of measuring vertical and horizontal angles, planes and curves). I wanted to know how they managed to align their imposing structures by the cardinal points and create such precise angles, internally and externally. I also wanted to know if they could measure latitude and longitude, as there is much debate and controversy in the area of ancient seafaring.

I was also intrigued by the concept of their measurements of the equinoxes and solstices, which were an extremely important aspect of their lives. Equinoxes are seen in the spring and autumn of the year and represent the rising of the sun and constellations at the same or "*equal*" point on the eastern horizon. Solstices mark the winter and summer high and low points on the eastern horizon of the sun and ecliptic constellations. For instance, in the summer solstice, the sun rises further north on the eastern horizon and has a higher elevation at noon. This is what causes the heat of summer. Conversely, during the winter solstice the sun rises further south on the eastern horizon and is at a lower elevation at noon. This is the cause of the colder temperatures of winter. The ecliptic constellations or the signs of the zodiac are at their lowest elevation at night at the summer solstice and at their highest elevation during the winter solstice. I knew that the ancients had named the constellations on the ecliptic, creating the signs of the zodiac and the astrological chart where each of the twelve constellations measures 30° totalling 360°. Evidence seems to show that they had knowledge of precession and possibly nutation, both of which are impossible to discover and calculate without the use of precise instruments, coupled with long term observations. To navigate using longitude requires the knowledge of local time against world time on the prime meridian. It is impossible to find longitude without this intimate knowledge of time. Therefore, I was curious about how they were able to measure time. For, to make oceanic voyages out of the sight of land requires this ability of time keeping by the navigator. All these tasks require a precise geometrical measuring instrument that can be demonstrated to work, otherwise these questions will remain a mystery and only supposition and speculation can exist. Academic institutions are extremely reluctant to acknowledge supposition and speculation as proof of existence. Rightly so, careers could be ruined and misinformation spread if supposition was relied upon too heavily. I am fortunate in this respect, in that I am an independent researcher and not bound by their rules.

The Experiment

I will explain to you now, how I embarked upon the discovery of this remarkable instrument. Part of my background is based in navigation as a qualified Royal Yachting Association Yacht Master. I started experimenting by attempting to find my latitude[xxi] at night. The latitude of a position is the angular distance, measured along the meridian, from 0° to 90° north or south of the equator. On this occasion however, I was pursuing this endeavour without use of a sextant or satellite navigation system. I wanted to reduce the potential tools to the very basic materials available to the ancient peoples based on archaeological proven discoveries. I invite you, the reader, to try the following methods that

I employed in my discovery, for yourself.

At first, I tried to sight the pole star using a short length of wood held to my eye at one end, with the other end pointing at Polaris (the current North Pole star in our epoch). I found the result to be very unsatisfactory due to the length of wood being unsupported and difficult to stabilise for accurate and steady viewing. I quickly realised that I would need a support to rest the sight on. I found that I needed an upright length of timber that was long enough to mount the cross bar on. The upright was to be at right angles to the cross bar so as to meet my eye whilst observing the star along the cross bar (*see the Patent illustrations*). The upright support had to be sufficiently long enough for the bottom end to act as a fulcrum when placed on the ground.

The result was a 'T' shaped contraption, approximately two metres in height. However, even though this was a more stable device, it was still not sufficient to measure the angle at a precise point on the Earth's surface due to the Earth's curvature and the parallax viewing angle of the pole star. A parallax is caused because of the immense distance of a viewed object. Parallax really means that there is no convergence of the angle to the object. In other words, the view is parallel no matter where one is positioned on the surface of the Earth. To overcome the problem of a parallax, another reference position has to be applied. This reference has to be constant. There is such a reference point that overcomes the problem of parallax and that is the centre of the Earth.

Ancient Plumbline Technology

Archaeologists are aware that the ancient Egyptians had used the plumbline as a method of measuring and weighing. The use of this method is seen at the weighing of the heart ceremony in the Egyptian Book of the Dead,[xxii] where weighing scales are employed for the judgement of the soul. The scales are operated by Anubis, who has placed the heart of the deceased on one side of the scale where it is balanced against the weight of a feather on the other. Anubis sets a plumbline at the crossbars of the scales to ascertain that they are perpendicular. A plumbline will always point directly to the centre of the Earth no matter where you are on the Earth's surface. This point of reference is a constant and never varies. I realised that if a plumbline was added to my sighting bar and fulcrum, I would always have an angle to work the parallax against. All that was required to find latitude was a scale measure of 90° for the plumbline to be read against. Therefore, by using basic geometry and the angles of a right angled triangle, it was theoretically possible to find the angle of the pole star, as it correlated to the plumb line when hung from the cross piece of the instrument that

I had made. I needed a round scale plate that would display 360° and fit on the centre of the cross bar with 0° at the fulcrum end of the upright.

It was then that I became inspired. I had recently been reading about the re-discovery of the Dixon relics found in 1872 in the north shaft of the Queen's Chamber, which is in the Pyramid of Khufu.[xxiii] One relic was a granite ball, which was thought to be a plumb bob (this granite ball is now on display at the British Museum). The strange discovery of these artefacts was brought to public attention by Robert Bauval and Adrian Gilbert in the book *"The Orion Mystery"*[xxiv]. Robert Bauval had made a reasonable comment in his book that a priest had hidden them behind the northern shaft of the Queen's chamber at some time in the past. Because of the discoveries of Rudolph Gantenbrink and his Uphaut Project, this is unlikely to be the case. The evidence from Mr Gantenbrink's robot cameras shows that the artefacts could only have been placed within the shaft at the time of construction.[xxv] To state otherwise would mean that there are further secret chambers within the pyramid that have not yet been discovered. No, it came as a shocking realisation, that these artefacts were as old as the pyramid itself and they were placed in their position by the original architect. This could indeed be the forgotten instrument of Thoth. Any researcher observing the results of Rudolph Gantenbrink's evidence can see that Waynman Dixon had damaged the relic at the top of the shaft, with an iron probing rod similar to that used by a chimney sweep. There is no doubt in my mind that he broke the complete instrument, which was already disassembled, dislodging part of it so that it fell to the bottom of the shaft nineteen meters further down from the original resting place, allowing Dixon to recover it.

Since I showed Robert Bauval the results of my research in March of 2001 at his home, he has embarked on a campaign to recover the missing piece of wood said to be a part of a scale ruler or measure. His aim is to have it carbon dated so as to determine its exact age, and date the construction of the pyramid once and for all. At the time of writing this book, Robert has located this missing part in Scotland.

Constructing the First Working Cross

I determined to see if this instrument could be constructed and used to the full potential that I suspected. A shopping trip was in order to purchase the material; two lengths of softwood, 50 millimetres by 25 millimetres thick and 3 meters in length. Some screws, nuts and bolts, and a builder's plumb line was all that was needed. I made the round plate out of a piece of plywood and using a simple school protractor and a ruler, extended the degrees to the circular outer edge. A hook was placed at the centre of the plate from which to hang the plumb line and bob, so that it could hang freely allowing the line to rest against the scale.

I completed the first instrument in early March 1998. I was completely astounded when it dawned on me what I had made. It appeared that I had made a replica of the Celtic Cross, or more likely, I was looking at the real Celtic Cross, a *working* cross and all the others in metal, stone and wood all over the world were replicas. Is it a coincidence then? Another coincidence in that I had independently discovered a completely new tool? A tool that looked the same as the ones ancients had made? Was it true that these objects in stone were an abstract idea from some befuddled past?

I tested it from my back garden one clear night and was able to read my latitude directly from the scale to half a degree. In other words, I was able to determine my position with this crude and basic instrument to within thirty miles on the Earth's surface. I was able to say thirty miles because one degree is equal to sixty nautical miles. I also was able to take levels in the surrounding countryside, to measure the heights of trees and work out inclinations.

A Unique Ability of the Cross

By turning the instrument round and using the top of the upright as a sight, I was able to measure the angle of pitch on roofs and all kinds of constructions. This sidereal measuring capability is unique. There is no other instrument that can achieve this and it is the main feature that makes this incredibly simple mathematical instrument so beautiful.

The cross is a physical manifestation of the mathematical solution.

The cross is capable of all spherical measurements, horizontally, vertically and on all planes.

It fulfils the mathematician's ideal. It is a bridge between the mental and the practical.

In this photograph, the author is demonstrating the use of the Celtic cross to measure the angle of the sun. The wheel is weighted and on a freewheeling axle. Because of the effects of gravity, the weight causes the bottom of the wheel to remain perpendicular regardless of the angle of the cross and the degrees are read from the scale around the outside as they fall in line with the centre of the upright. You will notice that the cross forms a pyramid shape as it is operated.

The cross is truly a cosmic instrument and therefore is the true definition of the Egyptian name of "Upuaut", the finder, or opener, of the way. This is why it is seen on Christian gravestones, it is an ancient symbol of an instrument designed to show the faithful how to find the way to God and Heaven. As a result of the unique feature of the cross (the capability of sidereal measurements), I am able to measure the declination of the constellations, stars, moon and planets from East to West, North to South, just as I believe the ancients had done long ago. I could tell the time with an almanac at night by sighting a star to measure its angle against the southern horizon.

The instrument could be used as a sundial in sunlight, by measuring the angle on the scale once the gnomon was aligned.

The Theory has been proven

My theory showing the cross as a navigational instrument has been reviewed and published by my peer group. This particular peer group, like me, has knowledge of navigation and seamanship. It took them two years to complete the review and they finally printed my work on finding latitude with the cross in the renowned and prestigious yachting magazine, Practical Boatowner in April 2001.[xxvi] The burning question was why had I not heard of this before, especially when it was so simple an instrument. I am known to be a level headed, practical and socially aware person, with a strong belief in justice, fairness and honesty, but something was seriously wrong. I am well read and such a possibility of an instrument such as this existing had never been recorded in anything that I had read. Why did the churches not seem to know of this tool? Why were all the teachers, the academics, the historians, and the archaeologists not aware of this simple instrument? In all the books on archeo-astronomy and ancient navigation, there are no clues to this type of instrument. How then, had I managed to not only rediscover it, but also be able to understand its usage and importance?

The above photograph shows the author measuring the angle of the sun against a Neolithic monolith near Oban on the West of Scotland. Some of you may see the simulacra showing two faces on the megalith.

Many of the stone Celtic crosses show the degrees marked out on the circular plate and incorporate the angles that are caused by precession and control the changing seasons. In other words the incorporation of the knowledge that there is a tilt of the planet of 23.4° in relation to the sun. I could only assume that the knowledge was originally a secret that had become lost in the fullness of time, as have many things. Despite the wonder of discovering this instrument in this form, even though it had been proven to work, I knew that the circular plate system was not capable of the accuracy required for measuring the altitudes and declination of planetary bodies sufficiently for astronomy, timekeeping or accurate navigation. Useful though it is for simple surveying, the scale is too small. It would need something of a much larger size, but a scale that was big enough to measure with sufficient accuracy would be too unwieldy, so it was back to the drawing board.

Why in this life, do we take everything for granted, without reason and investigation? We believe anything that is fed to us from an early age. The majority of us do not question our teachers; we believe what we are taught is the truth. It never even enters our mind that these people in authority, with whom we entrust so much, may not be teaching us the truth or perhaps they are victims too and do not know themselves. Is it a case of the blind leading the blind? Did dictators like Hitler exploit the old saying *"In the country of the blind, the one eyed man is King"*? Our powers of discrimination are supposed to be for survival and yet we are unable to discriminate between a religious icon and a mathematical instrument when it is constantly put in front of us.

I decided to continue on my journey of discovery and divulge the truth to whoever was listening or to whoever wanted to know. I carried out research on the Internet and only found a simple inclinometer made with a straw and a protractor for the instruction of children, this was the closest they came to my discovery. I researched all types of books and publications (See the bibliography) for any indication that I was not the only one who knew of this instrument's existence. I read through everything I could find on the Knights Templar as they were considered, in the middle ages, to be the masters of geometry. I visited many well-known Cathedrals in England, Scotland and France, spending a considerable length of time at Roslyn Chapel, near Edinburgh. After long nights of deliberation, I decided to make a patent application for the cross and pay for a full substantive examination in 1998. Not only had I rediscovered the working cross, I had actually been able to reconstruct several types proving that the cross actually worked. This ancient instrument was an incredible scientific tool and I felt it was and is my duty to reveal it to the world for the first time in over two thousand years.

To prove a theory, a model must be made and proved to work as in any scientific experiment. I have achieved this through the publication and granting of the UK Patent. The people who carry out the research in testing a patent are commercial experts in the field. It is their business to be thorough since they are a Government body.

Despite the substantive searches undertaken, the British Patent Office failed to come up with any findings of a similar instrument in the entire world. I received the results of the search and nothing similar had been found, although there were some clever applications using some of the independent attributes of the principles incorporated in my discovery. In other words, the overall concepts are known, but broken into disjointed applications and tools. None of them were whole or complete.

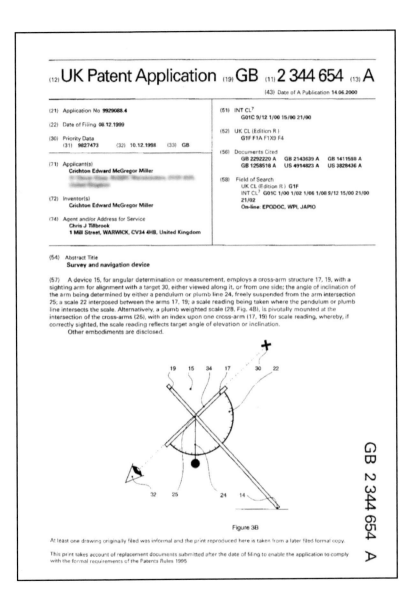

(12) UK Patent Application (19) GB (11) 2 344 654 (13) A

(43) Date of A Publication 14.06.2000

(21) Application No 9929088.4

(22) Date of Filing 08.12.1999

(30) Priority Data
(31) 9827473 (32) 10.12.1998 (33) GB

(71) Applicant(s)
Crichton Edward McGregor Miller

(72) Inventor(s)
Crichton Edward McGregor Miller

(74) Agent and/or Address for Service
Chris J Tillbrook
1 Mill Street, WARWICK, CV34 4HB, United Kingdom

(51) INT CL⁷
G01C 9/12 1/00 15/00 21/00

(52) UK CL (Edition R)
G1F F1A F1X9 F4

(56) Documents Cited
GB 2292220 A   GB 2143639 A   GB 1411588 A
GB 1258518 A   US 4914823 A   US 3828436 A

(58) Field of Search
UK CL (Edition R) G1F
INT CL⁷ G01C 1/00 1/02 1/06 1/08 9/12 15/00 21/00
21/02
On-line: EPODOC, WPI, JAPIO

(54) Abstract Title
Survey and navigation device

(57)   A device 15, for angular determination or measurement, employs a cross-arm structure 17, 19, with a sighting arm for alignment with a target 30, either viewed along it, or from one side; the angle of inclination of the arm being determined by either a pendulum or plumb line 24, freely suspended from the arm intersection 25; a scale 22 interposed between the arms 17, 19; a scale reading being taken where the pendulum or plumb line intersects the scale. Alternatively, a plumb weighted scale (28, Fig. 4B), is pivotally mounted at the intersection of the cross-arms (25), with an index upon one cross-arm (17, 19) for scale reading, whereby, if correctly sighted, the scale reading reflects target angle of elevation or inclination.
    Other embodiments are disclosed.

Figure 3B

At least one drawing originally filed was informal and the print reproduced here is taken from a later filed formal copy.

This print takes account of replacement documents submitted after the date of filing to enable the application to comply with the formal requirements of the Patents Rules 1995

GB 2 344 654 A

The Granting of a Patent for the Celtic Cross

The Patent was granted under UK Patent No: 2344887 in November 2000
Title: Surveying, navigation and astronomy instrument.

I HAD SUCCESSFULLY PATENTED THE Celtic CROSS!

The contents of the Patent, its use and construction methods may be studied in the appendix at the end of this book.

So, now that I have showed you how this amazing instrument is made and how it really works, is there any evidence other than stone images that indicate the cross was actually used by the ancients? Remember that a Patent is only granted because some of its attributes are not *"obvious"*. But these attributes are obvious, as you will see when I reveal to you what the ancient Priest Kings of the House of Aries hid in one of the "seven wonders of the ancient world." Let us first see some of the ancient mysteries that baffle the academics and historians. We will start with ancient architecture, in fact we will start with the architecture surrounding the Great Pyramid of Khufu. To solve that mystery of how the design of such an edifice of precise and massive proportions could have possibly been achieved. To do that requires an introduction to an even greater mystery of skilled and intelligent design as shown in the following article by Mark Foster.

THE TRIAL PASSAGES, A MESSAGE IN STONE?

Mark James Foster[xxvii]

Roughly 87 metres East of the Great Pyramid there lies a set of passages, hewn into the desert rock of the Giza Plateau. When they were first examined by Vyse and Perring in the 1840s they were thought to be passages from an abandoned pyramid or tomb, possibly even a fourth subsidiary pyramid of the Great Pyramid. [1] However, later on W.M. Flinders Petrie noticed that the passages seemed to be a very precise copy of the passages inside the Great Pyramid under the shadow of which they were cut.

He considered them to have been built before work on the Great Pyramid commenced, as an attempt to mock-up the layout of the internal passages in the edifice. Therefore they were named the Trial Passages.

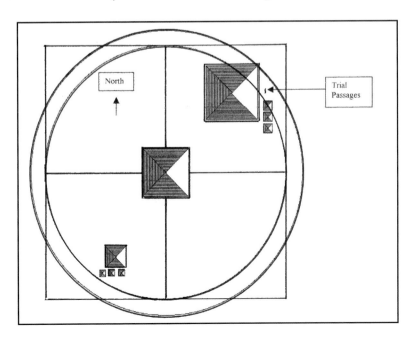

Is this really their purpose? Why have we not uncovered trial passages for any of the other 4th Dynasty pyramids or for a pyramid of any Dynasty for that matter? Egyptologists tell us that Khufu's father Snefru built both of the giant pyramids at Dashur before Khufu started work on the Great Pyramid. Both the Red Pyramid and Bent Pyramid contain complex chambers and passages yet there are no trial passages related to these buildings. Why have we also found no trial passages surrounding the Second and Third Pyramids at Giza?

What if the Trial Passages had some other purpose that has been overlooked?

Before we head down this route let's study the Passages in a bit more detail.

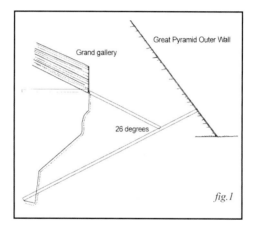

Take a look at Figures 1 and 2 and you will see the comparison. Figure 1 shows the passages inside the Great Pyramid whereas Figure 2 shows the Trial Passages.

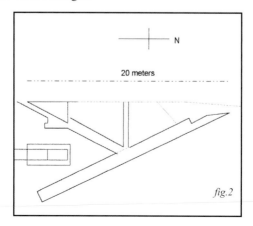

Mark Lehner also agrees with the striking similarity and comments;

> "*As Petrie recognised, these passages clearly are a kind of fore-shortened copy of the passages in the Great Pyramid*". [2]

Passage widths, heights and angles mirror the system of passages found inside the Great Pyramid. We have a Descending passage, an Ascending passage, the start of the Grand Gallery and the beginning of the Queen's Chamber passage. To further add weight to the idea that we are dealing with a replica of the inside of the Great Pyramid, the Ascending passage of the Trial Passages - where it meets the Descending passage - contracts as it does in the Great Pyramid as if it were ready to accept plug-blocks. No plug-blocks have been found in the trial passages yet the builders went to the trouble of adding this feature.

Furthermore, Flinders Petrie in his work The Pyramids and Temples of Gizeh [3] adds that there is also a passage that corresponds to the top of the well shaft found in the Grand Gallery in the Great Pyramid, however in this case the location is not in anyway identical to the position of this feature in the Great Pyramid. We will return to look at this passage in more detail later on.

Leaving the matter of the well shaft to one side, the remaining passages bear such a close relationship to those in the Great Pyramid that Petrie concluded: [4]

> "*The resemblance in all other respects is striking, even around the beginning of the Queen's Chamber Passage, and at the contraction to hold the plug-blocks in the Ascending passage of the Pyramid......The whole of these passages are very smoothly and truly cut, the mean differences in the dimensions being but little more than in the finely finished Pyramid masonry.*"

From this we can ascertain that a great deal of effort was expended to duplicate the internal passages of the Great Pyramid out on the desert floor, a stones throw from the pyramid itself.

Furthermore, the fact that the mean differences in the cut passages are little more than in the Great Pyramid itself (the passages of which are very highly praised by Petrie for their accuracy) [5] raise the likelihood that the two constructions were carried out by the same builders and that a high degree of accuracy was a requisite for both sets of passages.

Having looked at the layout and features of the Trial Passages let's now look at their purpose. I am going to put forward the hypothesis in this article that these passages were never intended as a trial for the internal passages of the Great Pyramid, rather that they had some other, hitherto unknown, purpose.

Firstly, I mentioned above that they seem to be unique in that we have found no traces of other trial passages connected with any other pyramid constructed in Egypt. If this was a standard practice in Ancient Egypt I would at least expect to find trial passages connected with the 4th Dynasty Dashur Pyramids or the Second Pyramid at Giza. Unless new evidence presents itself in the near future we will have to assume that the Trial Passages as we know them are unique and an anomaly that has nothing in common with any other Pyramid. I fail to see what context there is for Ancient Egyptians having used trial passages to construct their monuments. There is simply no evidence to suggest that this was the case. If the Great Pyramid was the first pyramid constructed by the Ancient Egyptian builders then I could be persuaded that they may have needed such a set of trial passages for their first construction. However, Egyptology tells us this is not so and that prior to the Great Pyramid several other impressive pyramids were constructed, all with complex internal passages and chambers and all without trial passages built first to mock-up their internal layout.

Secondly, I could accept the possibility that they might be trial passages if they had been constructed out of masonry. The fact is they were not, they were instead carved out of the bedrock of the Giza plateau whereas the majority of the corresponding passages were formed from masonry in the Great Pyramid. (Take a look at figure 2 again to establish what proportion of the Trial Passages are above ground in the Great Pyramid and therefore formed from masonry).

I fail to see the benefits of carving these passages out of hard rock when you are going to build in a completely different fashion when the time comes to start work on the Great Pyramid itself. Surely any experience gained would be negligible once the real construction began? I accept that a large part of the Great Pyramid's Descending passage was carved into the bedrock beneath the pyramid but if the architect was attempting to practice this art alone why did he then proceed to also carve passages into the bedrock that would be constructed from masonry in the Great Pyramid? If nothing else, the contraction of the Ascending Passage in the Trial Passages should make us sit up and take notice that something is not right here. Carving such a feature out of the bedrock would have required great effort and the experience gained would surely be virtually useless when repeating this feature with cut stones later in the Great Pyramid?

Because of these facts I surmise that these passages could never have functioned as trial passages. Is it simply the case that they have only been labelled as such merely because their true function has been overlooked?

So what possible function could they have had?

I certainly agree that the Trial Passages (I will continue to call them this for convenience despite my belief they do not assume this role!) were meant to represent the internal layout of the Great Pyramid. Furthermore, the builders went to great trouble to convince us that what we are looking at is a representation of the Great Pyramid - I'll draw your attention again to the contraction of the Ascending Passage - so that we could be left with no doubt in our minds. Think for a moment about a map or the key to a puzzle. With a key we are able to solve a puzzle precisely. What if the Trial Passages were such a key left for someone to solve the "puzzle" of the Great Pyramid at a later date? For such a key to work we would have to be left in no doubt as to what the key was pointing us to and this seems to fit what we know about the Trial Passages. Somebody went to a lot of trouble to carve the Trial Passages for apparently no practical reason - unless they were leaving us a very big clue carved in stone.

You will recall that history teaches us that it was the great Caliph Ma'mun who first discovered the granite plug-blocks stopping up the lower end of the Ascending passage like a huge series of corks in an even bigger bottle. In a recent article with Ralph Ellis we covered in detail this story and so I will not recount every point here but will instead refer you to that article. However, it is important for us to briefly recall some of the details.

Some time ago I was not happy with certain aspects of the tale of Ma'mun's discovery of the Ascending passage. We are told he was unable to locate the original entrance despite other sources that tell us that the entrance was known and open centuries before Ma'mun's time. We are then told that he arbitrarily began tunnelling into the pyramid at such a point that meant he struck the junction of the Ascending and Descending passages on his first attempt. The fact that this junction is offset by some 24 feet East of the centreline makes this all the more remarkable! My colleague Ralph Ellis then came up with the original and illuminating idea that Ma'mun's tunnel was actually dug from the inside out. *[6]*

If this is so we are left with one problem. If Ma'mun's tunnel was not a way into the pyramid but rather a way out then how did Ma'mun find the start of the Ascending Passage?

Firstly, I believe it has been demonstrated that the entrance to the Great Pyramid was accessible at this time and that Ma'mun would have ventured first down the Descending Passage into the Subterranean Chamber. *[7]* We are told that the beginning of the Ascending passage was covered with a hidden lintel that not only covered the plug-blocks behind but also meant that the passage was totally hidden from view and was indistinguishable from the other blocks forming the roof of the Descending passage at this point. Prior to the removal of this lintel entering the interior of the Great Pyramid would have been a very similar experience to entering the Red Pyramid at Dashur. You would have descended down one long passage before emerging into a chamber far below the entrance (there are differences due to the fact that the Red Pyramid's passage is built wholly into masonry but the effect is the same). In those days, The Ascending passage in the Great Pyramid would not have been visible at all and many early visitors to the Great Pyramid must have gone straight past it, completely ignorant of what lay above their heads!

Let us assume for a moment that many visitors were also aware of the Trial Passages to the East. How many puzzled explorers entered these corridors only to ponder why they were there? They led nowhere and there was no chamber at their conclusion. Instead they ended in a blank wall. Was Ma'mun the first to realise they bore a resemblance to the Great Pyramid's passages? Anyone measuring the width and height and angle of the entrance to the Great Pyramid and comparing them with the Trial passages would have seen a similarity. Was Ma'mun the first to do this and then notice that inside the Trial Passages there was a further passage leading up from the roof of the Descending passage? Did this make him wonder whether a similar passage existed inside the Great Pyramid?

I think it is highly possible that Ma'mun, fresh from this flash of inspiration in the Trial Passages goes back into the Great Pyramid and starts looking for a hidden passage in the roof of the Descending Passage. Without knowing it he has taken the builder's clue and has started the process of unravelling the puzzle.

At this point we have to consider how Ma'mun came to find the exact location of the hidden Ascending passage. Peter Lemesurier in his book The Great Pyramid Decoded raises the interesting point that a pair of scored lines carved into the walls of the Great Pyramid's Descending passage seem to bear a similarity with a sloped face found at the top of the Trial Passages' Descending Passage. *[8]* Looking at both these features they do indeed seem to indicate a precise point in both sets of passages.

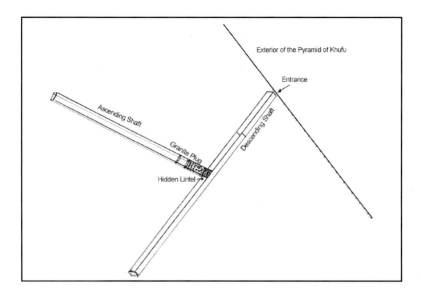

It occurred to me that it would have been possible to compare the distance between this feature and the beginning of each Descending passage to work out what scale would be needed to calculate the position of the start of the Ascending passage. The problems I have encountered with these calculations is that the ground around the start of the Descending passage of the Trial Passages is very worn and it is difficult to ascertain where the passage originally started. However, if it originally started at the height of the seemingly levelled ground above the passages as a whole, I have estimated that this would have enabled Ma'mun to have identified the exact spot on the roof of the Descending passage to within a few inches. This would have been more than accurate enough bearing in mind that the hidden lintel would have resembled one of the large roofing stones so being correct to within a few inches would have highlighted which stone was actually the hidden lintel and not a roofing slab.

Of course he may not have taken this into account and may just have tried examining all of the roofing blocks one by one until he located the right one! The truth is we will never know but it is very interesting to note that it could well have been possible for him to locate the entrance in this manner if he had studied the Trial Passages closely enough and made the connection between the scored lines and the flat surface of rock inclined at the same angle at the head of the Trial Passages. To further highlight this point it must be mentioned that the purpose of the scored lines within the Great Pyramid has yet to be argued conclusively so I find it highly likely that they are there for just this purpose, as a marker, a point from which measurements should be taken when compared to the Trial Passages for the sole purpose of locating the hidden Ascending passage.

Locating the hidden lintel Ma'mun then removes it and expecting to see an Ascending passage he is suddenly surprised to find that it is indeed there but travels only a few inches before being plugged by huge granite stones!

The fact of plugging the Ascending Passage seems to have been designed to intrigue and pique our curiosity even more. First the builders leave us a clue in the form of the Trial Passages. Once this clue is deciphered and the lintel removed Ma'mun finds to his dismay that the passage is plugged! What would anyone's reaction be in that instance? Anyone who tells me they would walk away and leave the plugs in place is a liar! Ma'mun did what any of us would have done, he tried desperately to get past them. After a few vain attempts at chiselling through the granite plugs themselves proved ineffective he ordered that they be dug around and in doing so he revealed the Ascending Passage. What happened next is well documented in the article co-authored by Ralph Ellis and myself and does not concern us here. Whatever was found in the upper chambers is an important but separate issue here. What is vital for our discussion here is that the plugs had been breached and modern mankind had progressed to the next stage in the puzzle. Ma'mun had done his part, he had taken us past the first stage of the riddle. The next part of the puzzle was to remain a secret for nearly 1,000 years until a man called Waynman Dixon started exploring the Great Pyramid.

At this point I would just like to consider the point of just why the builders would have left such a key for us. Firstly, readers of my work will have appreciated that I consider it highly unlikely that the Great Pyramid was merely a tomb. Whether it was ever the tomb of Khufu is debatable but I do believe from my own research that it is highly probable that the Great Pyramid was standing before Khufu's reign and had a purpose other than being a tomb. For the record, I do believe that the Pyramids of Giza were constructed by Egyptians, it is just that I attribute their civilisation a greater period of time than history currently affords it.

Therefore, leaving us a map of the insides of the Great Pyramid becomes slightly more interesting. The question now turns to why?

To my mind the Trial Passages are simply a pun carved in stone. Again this is not out of context with what we know of the culture of Ancient Egypt.

Scattered throughout Egyptian Texts we come across a variety of puns that were deliberately placed in the spells and lines of The Book of The Dead and other funerary writings. The Egyptians seemed to have a delight in such puns and great pains were taken to place them for dramatic effect.

One of countless examples is pointed out by Dr. Ogden Goelet in the Papyrus of Ani, Chapter 147 (speaking of Osiris);[9]

> *"The one purified by your own efflux* (setau) *against* (r) *which the name of Rosetau was made."*

By noting that the word Rosetau sounds much like the phrase 'against the efflux,' the fluids issuing from Osiris's body were thus made innocuous.

While this example is taken from a text dating to circa 1250 BC there are numerous examples all throughout Egyptian literature including the Pyramid Texts inscribed onto the walls of the 5th Dynasty Pyramids of Saqqara so it is evident that this kind of word play and punning was prevalent right throughout Egyptian history. While some of these puns were cleverly designed to enhance the power of the particular line, as above, others were much more simply a play on words for everyday objects, superimposing the meaning of one onto another for dramatic and sometimes humorous effect.

So the question is could they have incorporated such puns into their buildings? I think it is certainly within the realms of probability if the Egyptian culture used them this extensively in their funerary writings.

It would appear that this was indeed the case as the Trial Passages are not the only example of such a pun being left in stone at Giza.

I mentioned it would be almost another 1,000 years or so before the next clue was found. Waynman Dixon began his survey of the inside of the Great Pyramid with his brother in the year 1872. It is well documented that there were at that time shafts present in the King's chamber but none in the Queen's chamber below. Then Dixon looking at the shafts in the upper King's chamber wondered if there might be similar shafts hidden in the Queen's chamber. Measuring the position of the shafts in the upper chamber he estimated where they might be found in the Queen's chamber. Asking his man-at-work to get to work with a chisel two shafts were suddenly revealed.

Amazingly, the shafts leading from the Queen's chamber were not cut through into the chamber itself but instead were hidden some eight inches behind the walls! Furthermore we know now that the Southern shaft definitely does not make it to the outside of the Pyramid and it is very likely to be the same with the Northern shaft (this shaft has not been explored fully at the time of writing but there is certainly no evidence for the shaft emerging on the outside of the pyramid's face). Could it be the builders had left another pun for us to work out here? I certainly feel it is in keeping with the mind of the builders. The fact that the Southern shaft has now been explored only highlights the point. Gantenbrink's discovery of what appears to be a *door* or portcullis slab at the end of the shaft only adds to the mystery. *[10]* Whatever the so-called *door* turns out to be it is evident there is *something* there, something we would never have found had Dixon not solved this particular riddle.

I don't think it is irrelevant that the shafts themselves are only eight inches square. This speaks volumes to me concerning the mind of the builders. What better way of frustrating and intriguing the mind of the explorer than to build a passage that no man could ever ascend! Dixon tried exploring the shaft and failed, losing part of his iron rods in the process (they can still be seen stuck around a bend in the Northern Queen's shaft). It was only after Rudolf Gantenbrink's exploration by robot that we realised there was something at the end of the Southern passage.

Yet still the pun of the builders is at work because even though we now know there is a *door*, portcullis slab or something else at the end of the passage we are unable to get up there and open it. We know there is something of importance there because just before we reach the door the passage is suddenly lined with fine Turah limestone, reserved solely for chambers and casing in the Great Pyramid. I interpret this as the builders attempting to show us that there are indeed further chambers high up in the Great Pyramid . But what is the point of showing us there is such a chamber only to then highlight that there is no way we can get up to it?

Unless there is another way up?

Could the point here be that having shown us there is another chamber we have to find another way to reach it? Another passage that at the present time is unknown and hidden from us.

To solve this we must try and think the way the builders would have. I sat and thought about this for a while then realised it was time to go back to the original key, the Trial Passages, and look to see if anything had been missed. I was instantly surprised by what I found. It became clear to me that the purpose of the Trial Passages is to focus attention on one particular part of the Great Pyramid only, the junction of the Ascending with the Descending passage. This becomes clearer the more you look at it. Everything centres on this feature, and other features such as the Grand Gallery are only sketched in so we that we are able to work out that we are indeed being shown a map of the internal passages of the Great Pyramid. Features like the Subterranean chamber are not shown at all, neither is the Queen's chamber. Furthermore, if you examine it, it is evident that this is the only portion of the whole set of passages that is completely to scale, other features are foreshortened but the junction does not suffer from this problem of scale, it is there complete. It is evident from this that we are being forced to look long and hard at the junction of the Ascending with the Descending for some very important reason.

Therefore, taking this lead from the passages themselves I studied this junction until I realised that I had been staring at the solution all along without actually seeing it. The answer could be there right in front us screaming for attention but we have failed to see it up until now.

There is a passage in the Trial Passages that does not appear in the Great Pyramid. Or put another way, a passage that *is* *there* but we simply haven't uncovered yet. Remember that before Ma'mun discovered the Ascending passage nobody knew it was there! Could it be that there is another passage we are walking past everyday, one that leads up into the upper reaches of the pyramid?

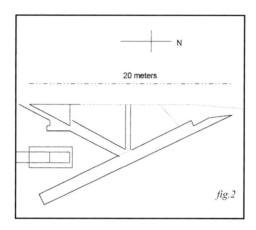

fig.2

Take another look at Figure 2, reproduced again here. You will notice the vertical passage slightly to the North of the start of the Ascending passage. We have yet to find a passage inside the Great Pyramid that corresponds to this feature in the Trial Passages.

I believe that this fact has been overlooked because it has been assumed it represents the Well Shaft. Petrie certainly believed this is what it represented, saying; [11]

> *"The vertical shaft here is only analogous in size and not in position, to the well in the Pyramid gallery; and it is the only feature which is not an exact copy of the Great Pyramid passages, <u>as far as we know them.</u>"* [Petrie's own emphasis.]

I have to take issue here with Petrie and while I have the greatest respect for his work I have to conclude he has made a serious error here in comparing this shaft to the Well Shaft. If we look at the position of the Well Shaft in the Great Pyramid it can be seen that it begins in the Descending passage and after many twists and turns emerges at the base of the Grand Gallery, arriving beneath a loose stone that formed part of the West ramp of the Grand Gallery. Look again at the shaft in the Trial Passages. How can this passage be comparable with this feature in the Great Pyramid? The position of the two shafts is not even remotely close. A vertical passage in such a place in the Great Pyramid would be nowhere near emerging at the base of the Grand Gallery. To my mind to state that the two are analogous is stretching a point too far. I can only conclude that they relate to separate features and can in no way be thought to be the same feature.

However, in reading Petrie's words again I am forced to consider the fact that he actually suspected himself that they were not the same feature. What does he mean by the words "*the only feature, which is not an exact copy of the Great Pyramid passages, as far as we know them*"? Did he suspect that one day it might be revealed that this feature could indeed be found in the Great Pyramid? We will never know but what we can do is take his lead and think about what he is saying.

Having surmised that it is highly likely that this is not actually the Well Shaft we have to take onboard the message the Trial Passages seem to be pointing out to us here.

There is another passage in the Great Pyramid, one we have not uncovered yet.

Furthermore it leads up vertically into the body of the pyramid from the junction of the Ascending passage with the Descending. If this passage was indeed to be found inside the Great Pyramid where would it lead? It does not necessarily have to remain a vertical passage. It could quite simply lead anywhere, it could develop into a separate passage system and lead to new chambers. The truth is it is impossible to tell. The Trial Passages leave this question unanswered. The beginning of the vertical shaft only is represented in the Trial Passages in much the same way as the beginning of the Grand Gallery. I believe this is so for the reasons outlined above, it is the junction of these passages only that the Trial Passages are attempting to bring to our attention. They seem to want us to make a breakthrough and discover the passage. Where it leads will become evident at that time.

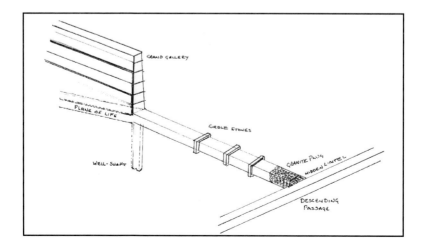

So does this passage really exist in the Great Pyramid and if so why has it been missed?

I believe it is highly probable it really is there. The reason it has remained hidden for so long is due in part to the ingenuity of the architect. Look again at the plug-blocks in the Great Pyramid. Now refer back to the vertical passage in the Trial Passages. It is clear that the plugs start at exactly the spot where we would expect to find the beginning of the vertical passage and do not cover the spot where we would expect to find the entrance to the vertical passage. Why then can we not see the passage?

I believe this passage was designed not to be found by accident. Therefore I believe it is hidden in almost a similar way to the shafts found in the Queen's chamber. Remember that those shafts were always there, hidden behind the wall. It was only the foresight of Dixon that revealed them. I do not believe it is unreasonable to suppose the vertical shaft really is there, exactly where the Trial passages show it to be, it is simply a case that the lower end has been made to look identical to the roof of the junction between the Ascending and Descending passages. Again, this is not an unreasonable suggestion given that we have already encountered a similar method of hiding a passage within the Great Pyramid itself (the Queen's shafts).

Furthermore the fact that the plugs slope up at the angle of the Ascending passage indicate that the prime purpose is to block that passage only. Using granite for the first three plugs (Ma'mun found further limestone plugs blocking the Ascending passage which he was able to remove) seems also to be part of the design because it ensures anybody uncovering them would dig around them rather than try to remove them. Had they been limestone like the rear plugs they would have been easier to remove. This ensures that the correct sequence is followed by anyone following the clues hidden in the Trial Passages. Therefore Ma'mun dug around the blocks in the direction of the Ascending passage and so came to discover the Grand Gallery and the upper chambers without disturbing the vertical passage. Putting the plugs so close to the vertical passage means if there is a lintel hidden above the plugs it is protected by the plugs.

So, how much stone prevents us from gaining access to this new passage? Could it be as little as eight inches as in the case of the Queen's chamber. It is an exciting prospect.

I hinted earlier at where such a passage would lead us to. If there were chambers and other passages leading from this vertical passage it would follow that these may impact upon the passages and chambers we already know about. A few final pieces of evidence present themselves here.

Firstly, some way up the Ascending passage the construction of the passage changes in several places and it has been revealed that there are three Girdle stones fixed in place. These are obviously designed to strengthen the passage at this point. If we look at a diagram of the Ascending passage it is obvious they only occur in one section of the passage. Why is only this point of the passage strengthened? Why was it not necessary to place girdle stones along the length of the whole passage? There is no evidence of similar girdle stones in the upper part of the Descending passage or in the corridor leading to the Queen's chamber so why here at this precise point?

Could it be that it was necessary to strengthen the corridor at this point because there was something above this part of the passage? I think this is a very plausible suggestion. A chamber directly above the Ascending passage would cause the weight of the surrounding core blocks to be shifted unevenly around it, possibly to such an extent that strengthening blocks were needed to keep the passage below intact. It seems very possible to me that the presence of girdle stones at this particular spot indicates the position of an, as yet, unknown chamber. The position of the girdle stones is very close to my suggested vertical passage as the diagram on the left shows. I don't think it is outside the realms of possibility to suggest the two features share a connection.

Our second piece of evidence comes from Rudolf Gantenbrink's survey of the shafts found in both the King and Queen chambers. In particular, from anomalies he found some way up both the Southern shafts. Here is an extract first from his exploration of the Queen's Southern shaft; [12]

> *"At the beginning of Block No. 26, a large section of the floor has broken away. This is the worst damage we observed anywhere in the shaft sequences so far investigated. At this point, however, the pressure on the shaft amounts to only one-third of the maximum value. Near the Queen's Chamber, 115 meters of pyramid material press downward on the shaft. But only 35 meters of material press down on this spot, where we observe the greatest shaft damage. This highly unusual finding can have resulted only from one of two possible causes:*
>
> *1. Extremely inept construction work below Block No. 25 and 26. It must be remembered, however, that it is this final section of the shaft*

which otherwise displays the highest quality workmanship observed anywhere in the shafts system.

2. The existence of an as yet undiscovered structure below or above this shaft section. Such a structure could produce a pressure peak, which could in turn focus considerable additional force on the shaft and possibly cause the observed damage."

Furthermore, he later finds more evidence further up this shaft that seems to reinforce the idea that there might be an undiscovered structure in the pyramid somewhere in this vicinity when cutting marks are found on the floor of certain blocks forming the shaft; [*13*]

"Based on the grooves found in the shaft, we can assume that, before their insertion as floor slabs, these blocks served as a base for the cutting of precision joints.

This gives rise to a crucial question: exactly which precision joints were cut here?

The shaft blocks themselves were only dressed with the chisel. We observed ample evidence of this in the Caviglia Tunnel, on the lower sides of the shaft blocks, as well as at several sites of block displacement, which exposed the abutting edges. At the upper southern shaft outlet, both outer sides of the blocks are visible. These, too, were worked only with the chisel. Thus, as 9 of a total of 10 surfaces of a shaft block were definitely chiselled, we can well assume that the shafts were constructed without recourse to sawing.

The pyramid's nearest casing stones, which lie 19 meters distant from this spot in the shaft, were cut in their final position. We know this because the stones located directly beneath the casing stone display cutting grooves.

The pyramid's corridor and chamber system, which also displays precise, cut joints, had been completed long before this shaft construction level was reached.

Taken together, these findings constitute a compelling case for a possible, as yet undiscovered structure - for which precision joints where made - in this upper region of the southern, Queen's Chamber shaft."

To add yet more weight to this hypothesis are his findings in the King's Southern Shaft, at a point directly above the anomalies found in the lower, Queen's Southern Shaft; *[14]*

> *"Between Block No. 15 and 16 we discovered a vertical joint. In the shafts such joints, which have a distinct static function, otherwise occur only proximate to the chambers.*
>
> *It is a complete anomaly to find a vertical joint fully isolated in the nucleus of the pyramid. Since it requires much greater effort to shape and fit the blocks in such an arrangement, we can assume that the builders must have had significant structural justification for going to the trouble of deflecting forces into the horizontal plane.*
>
> *This vertical joint is located about 12 meters above a point in the lower southern shaft, which is subject to extraordinary static influences. The overall statics in this area seem to differ from those in the other shaft segments. For a construction engineer this is a significant clue to the possible existence of an as yet undiscovered structure in the vicinity of these static anomalies."*

This coupled with the evidence I have put forward concerning the Girdle Stones in the Ascending passage I believe shows we should expect to find further chambers within the Great Pyramid, possibly even a whole new set of passages and chambers leading from my proposed vertical passage. As for the question of whether this new set of passages leads ultimately to the other side of the door found at the end of the Queen's shaft, I am afraid I will have to leave that question to the first person who enters the vertical passage! I will simply say that it is not beyond the realms of possibility bearing in mind it is highly likely there is something on the other side of the door and we cannot possibly reach it via an eight inch passage.

Thankfully we live in an age where it is no longer necessary to destroy large parts of the Great Pyramid's internal passages in order to answer some of these questions. Had Howard Vyse stumbled upon what I have revealed he would have undoubtedly had his men dynamite the junction of the Ascending passage with the Descending to establish what lay beyond, simply because they were the only means available to him at the time. Today though we have developed less intrusive methods and I believe it should not be beyond modern technology to confirm whether this vertical passage is really there or not.

For my own part I trust the message of the Trial Passages and believe there is a very strong likelihood that this passage could be found if we only take the trouble just to look. Exactly what we will find and why the builders of the Great Pyramid went to such lengths to lead us on this voyage of discovery are questions I cannot answer right now. If the passage exists as I think it does then all will become clear once we follow where it leads us.

I am confident that even if we do not heed the clue at this moment in time somebody will at a later date. Maybe, as it was in Ma'mun's time, it will be another 1000 years before someone takes the next step.

© Copyright Mark James Foster 2001

I would like to thank the following without whose help this article would not have appeared; Ralph Ellis, for his many discussions regarding Ma'mun and his exploration of the Great Pyramid and for also helping me see that the possible vertical passage in the Great Pyramid may be hidden in a similar fashion to the Queen's shafts. Simon Cox, for his research into Vyse's diaries, for supplying source material, for help with the original draft and offering numerous suggestions as well as encouraging me to finish the article! Greg Taylor, for proof reading and correcting my poor grammar!

NOTES

1. Vyse, H, Operations Carried on at the Pyramids of Gizeh (3 vol.) 1840-1842, London
2. Lehner, M, The Pyramid Tomb of Hetep-heres and the Satellite Pyramid of Khufu, DAI
3. Petrie, W.M.F, The Pyramids and Temples of Gizeh, 1990, HMM
4. Ibid.
5. Ibid.
6. Ellis, R & Foster, M.J, Tunnel Vision
7. Ibid.
8. Lemesurier, P, The Great Pyramid Decoded, 1977, Element
9. Faulkner, R & Dr. Goelet, O, The Egyptian Book of The Dead, The Complete Papyrus of Ani, 1994, Chronicle
10. The Upuaut Project - A Report by Rudolf Gantenbrink http://www.Khufu.org/
11. Petrie, W.M.F, The Pyramids and Temples of Gizeh, 1990, HMM
12. The Upuaut Project - A Report by Rudolf Gantenbrink http://www.cheops.org/
13. Ibid.
14. Ibid.

This article first appeared on Mark Foster's web site - www.rosetau.com
© 2001 Mark James Foster.

What I put to you now is a question of reason and logic. When even a simple project such as the building of a modern housing estate requires complex surveying and planning with sophisticated instruments and mathematics prior to building, how could this degree of craft, skill and knowledge, as expressed in the complex of Giza, possibly be achieved with no instrument for measuring inclines?

Well, I believe there was such an instrument and I have discovered one in the Great Pyramid of Khufu.

THE CROSS IN THE PYRAMID

I reverted to the report on the relics found in the Pyramid of Khufu.[xxviii]

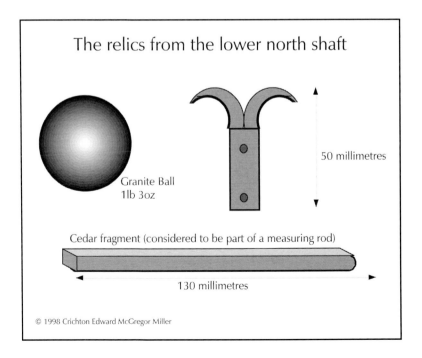

The relics from the lower north shaft

Granite Ball
1lb 3oz

50 millimetres

Cedar fragment (considered to be part of a measuring rod)

130 millimetres

© 1998 Crichton Edward McGregor Miller

I had initially dismissed the metal "hook" as only being a decorative method of attaching the plumb line to the cross bar. From a decorative point of view, it is a fair representation of the ancient symbol for the zodiac sign of Aries. I have since discovered through experimentation that it has another, much more significant purpose. I took great interest in the section of the report stating that there was part of a measuring rod amongst the finds. I then incorporated this rod into the construction of a more complex version of my instrument. I had to adapt a rule that would work, as the construction of a measure divided into degrees would be too difficult to manufacture accurately enough for the first experiment. Some improvisation was required so I adapted a one-meter rule for the purpose. Using only 90 centimetres of the rule, I created a right-angled triangle between the rule, the upright and the end of a cross bar (*this design may be seen in the patent drawings*). For this experiment each centimetre marked out on the rule is equal to one degree of arc. Again, the experiment worked and showed my latitude about six miles from my position according to the Global Position Satellite receiver that I checked it with. Later, I was to develop a type of scale rule that worked more efficiently.

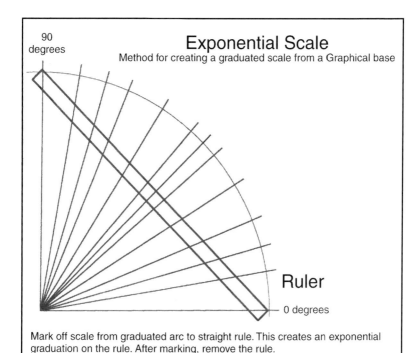

90 degrees

Exponential Scale
Method for creating a graduated scale from a Graphical base

Ruler
0 degrees

Mark off scale from graduated arc to straight rule. This creates an exponential graduation on the rule. After marking, remove the rule.

The greatest errors are in angles over 60 degrees. As well as below 30 degrees.

I went on and constructed an exponential scale. An exponential scale is one where the degrees on a circle are extended from its centre and laid across a straight rule. As the scale is viewed from left to right or vice versa, the units of measurement become wider from the centre of the rule. In navigation, one degree is equal to sixty nautical miles and one minute of arc is equal to one nautical mile.

The Gods of Time become apparent again as distance becomes associated with time. An ancient navigator, who was equipped with an instrument which was fitted with a graduated rule of around one meter in length, would only have to travel six miles from one observation point to the next, to discover the circumference of the Earth through the north, south axis. This would be achieved by sighting the North Star or stars of the time and observing the resulting angle. A short journey north or south on the same meridian and the taking of a further observation using the same pole star would show a difference of one minute of arc. To reduce this to its simplest form of measurement, the distance of each pace of the walking observer could be measured as, say, "paces" or feet or yards. These are composed into larger units of miles and the system of measurement is simply multiplied by 360 to find the distance. There may have been error of course, as there is no evidence that the ancients understood that the Earth was an oblate spheroid. The Earth as an oblate spheroid means that it is flattened at the poles.

THE NAUTICAL CONNECTION

Consequently, latitude and longitude could, in theory, be discovered with a surprisingly high degree of accuracy with the cross. In the case of longitude an accurate knowledge of the local time zone would be required. This is covered later in the chapter about the finding of longitude. To find latitude with the use of a plumbline would logically lead to the rational conclusion that the ancients knew the Earth was a sphere. There could be no other conclusion possible because of the nature of the constancy of the plumbline in combination with the effects of gravity. The finding of latitude is only possible on a sphere, so as soon as the academics make any reference to this as part of their theories or proofs, they must acknowledge that the ancient Egyptians knew the Earth was round.

The relevance of this is the latitude of the Giza complex. It appears to me to have been strategically placed in that position, 6.6 degrees north of the maximum altitude of the zodiac stars and the sun on the ecliptic plane (*see the description of the ecliptic in the chapter on the zodiac*). There is a considerable connection between the pyramids at Giza and an ancient nautical influence. Boat pits have been discovered around the pyramids and large high prowed ships have been buried within them. One of these ships, built using Lebanese cedar has been recovered and is now housed in a museum within the Giza complex. It was buried shortly after the death of Khufu. It has been associated with the funerary process only and therefore not given sufficient attention by scientists with knowledge of the techniques of seamanship. I do not think that this gigantic ship, that is over forty-three meters long, was used to go to sea, but it is copied from a style that could have been used for such purposes. It would have been capable of carrying a crew of over one hundred sailors. Its prow and stern are designed to part ocean waves of an unlikely size to be encountered on the Nile and it is over four thousand years old.*xxix*

There are some of the best examples of seagoing ships to be seen on the mortuary temple of Queen Hatshepsut at Deir-el-Bahari at Thebes.*xxx* The friezes date to the New Kingdom period, 1570 to 1070 BC, they are over twenty meters long and over five meters wide. The sail plan was very efficient and capable of upwind work or using the power of the wind to move the vessel towards the wind. A science like this is not discovered overnight, it is a question of great experience and practical use collected over a great deal of time and also requires an understanding of geometry. Sailing ships are sleek machines working in close harmony with the elements and promoting admiration in all who see them, they are by far the most beautiful of mankind's inventions. They are capable of travelling the world, providing that the ship's master knows where he is. In the case of ocean navigation, to be accurate to within six miles of a position is quite sufficient, as it would bring the observer within the visible limitations of horizon, so that pilotage could be used.

Pilotage is the method employed by a navigator when within sight of land. The art is to plot the ship's position through the observation of visible landmarks and in modern times, flashing lights and buoys.*xxxi* Six miles is quite accurate for mariners when making landfall, as when coupled with the height of the observer above the level of the sea, it is within the horizon. The cross and plumbline is capable of that accuracy and more.

In terms of accuracy, this distance is less than the ten miles of error in the first famous navigation of the Atlantic using a watch.*xxxii* Sir Isaac Newton said at the time of this great endeavour, that in his opinion, finding longitude would be better pursued through astronomy. It has been said that Sir Isaac Newton was a Gnostic and practised Alchemy, and as it has also been stated that Newton discovered his measurements of lunisolar precession at Giza, it can be understood why his views may have been influenced.*xxxiii*

The pyramids at Giza and their connections with a maritime heritage now start to form a pattern in the mind of a serious navigator. It is obvious that if you wish to navigate out of the sight of land, you must find your position. Make no mistake about this problem, it is essential to overcome or you will be ship-wrecked. The ships mentioned earlier and those that made recorded journeys around the Mediterranean Sea and the Red Sea thousands of years ago would have been out of the sight of land most of the time as no Captain will sail close to a lee shore or risk running aground. That's why, as a sailor and navigator, I find it laughable that certain academics believe that the Egyptians and Phoenicians thought the world was flat. I would love to take them to sea in a small sailing craft and show them that it's impossible to be fooled into thinking the Earth is flat. Firstly, you can see the curve of the horizon and secondly, you immediately notice that things disappear over the horizon as you leave them behind, and rise above the horizon as you approach. Landlubbers should not judge sailing, they are not equipped properly for this practical art. This ancient ability to navigate could be seen as a form of magic and it certainly formed the basis of the knowledge of astronomy, astrology and geometry.

A Great Nautical Treasure

I propose that the Great Pyramid was surveyed before construction by using a cross and plumbline which was the navigator's ultimate tool and magic staff of power. That it was sailors who had this knowledge from which all civilisations past and present were formed. I also believe that the relics discovered in the North shaft of the Queen's Chamber in the Great Pyramid of Khufu by Waynman Dixon and Dr Grant in 1872, may have been one of the greatest Egyptian treasures to have been discovered. At this time, this great treasure is the only preserved and physical manifestation of the ancient working cross to be found to date. Understanding these relics and their capability will open up all kinds of avenues of new research into the ancient past. Charles Piazzi Smyth, Astronomer Royal for Scotland, in his 1878 book "*The Great Pyramid*" recorded the relics found by Dixon and Grant. They had been discovered in the hermetically sealed north shaft broken into by Bill Grundy under the direction of Waynman Dixon and were sent to Piazzi Smyth in a cigar box where they were recorded in his diary, with accompanying drawings and sketches.

The subsequent loss of these relics was the subject of an extensive investigation in 1993 conducted by Robert Bauval. Robert could not have achieved this without the assistance of Dr. Mary Bruck and the late Professor I.E.S. Edwards (Dr. Edwards was the keeper of Egyptian antiquities at the British Museum as well as the acknowledged foremost expert on the pyramids). Dr Mary Bruck was a lecturer in astronomy at Edinburgh University and the wife of Professor Herman Bruck. Astronomer Royal for Scotland 1957-1975. The events are documented in the epilogue of Bauval's book, '*The Orion Mystery*'.

In his book, Robert Bauval mentions several times that one of these relics, the bronze hook, was probably a form of Pesh-en-Kef (see Bauval and Gilbert '*The Orion Mystery*') and also "a sighting device for stellar alignments". Professor I.E.S. Edwards also supported this view. He also suggested, like the Czech Egyptologist and archeo-astronomer Zbynek Zaba before him, "that the Pesh-en-Kef instrument, fixed on a wooden piece and in conjunction with a plumb-bob, was used to align the pyramid to the polestars". Furthermore, he suggested that "it seemed very likely that a priest placed the ritualistic tools inside the northern shaft from the other side of the wall of the Queen's Chamber". This prompted him to state further in '*The Orion Mystery*' that "we cannot help wondering if these ancient relics (were) indeed, perhaps the very sighting instruments that were used to align the Great Pyramid to the stars".

What all these learned gentlemen missed was the incredible complexity of the cross in its full glory and power. They had missed the ability of the cross to perform remarkable mathematical tasks and skills. They all overlooked the great architect's delicate and ingenious instrument, skill and depth of knowledge. No other type of instrument could have been more efficiently employed in the design of the mightiest construction of all time, The Great Pyramid of Khufu. I believe that the Giza complex was the greatest clock in the ancient world and that this edifice was the great throne of the Gods of Time.

This cross, the answer to one of the greatest mysteries ever presented to Man, was broken and damaged in 1872 by a clumsy attempt to shove an iron rod up the shaft in the manner of a chimney sweep or drain cleaner. The broken pieces fell to the foot of the shaft and were recovered by Waynman Dixon and subsequently transported in an undignified cigar box and lost for over a hundred years. It is only now that it is, at last, attracting the interest that it deserves.

The detailed drawings and notes of the Dixon artefacts from Entry 26, November 1872 in Piazzi Smyth's diary are now kept by the Royal Society of Edinburgh. Robert Bauval obtained permission to publish them in '*The Orion*

Mystery' from the secretary Dr W Duncan. On the matter of the ball, it was weighed by Piazzi Smyth and noted to be 8,325 grains or 1lb 3 oz (approx. 0.850kg). This would work effectively as a pendulum for the proposed instrument as the weight would settle the oscillation of the line quite effectively.

In an interview with Dr Carol Andrews of the Egyptian Antiquities Department of the British Museum, held by Robert Bauval on the matter of the Pesh-en-Kef, Dr Andrews felt that it was unlikely to be a Pesh-en-Kef instrument as no Pesh-en-Kef instrument was known to have been in existence prior to the eighteenth dynasty. Dr Andrews also favoured a stellar alignment instrument. Those Egyptologists who have studied the Pesh-en-Kef, may dispute Dr Andrews' view since they feel that it was an instrument that was only designed as a form of adze that was used in the *"opening of the mouth ceremony"*.[xxxiv] However, the concept of the ceremony may not yet be fully understood.

There are some who may also feel that the granite ball was not a plumb bob as it does not resemble the examples of plumb bobs both illustrated and seen in ancient artefacts. However, in my opinion, I would consider that the Dixon artefacts are of earlier origin than those plumb bobs we use for reference purposes. This may be seen in the photographs of the mysterious Neolithic stone balls found in certain areas of Scotland and covered in the chapter about Kilmartin toward the end of this book.

The items found by Waynman Dixon in the 1872 discovery, sent to Edinburgh in a cigar box and recorded by Piazzi Smyth, consisted of:

- A slat or rod of cedar wood about 13 centimetres long (part of a measuring rod?).

- A granite ball weighing 1lb 3 ounces.

- A bronze/copper hook type of instrument, 5 centimetres long, with a part of a wooden handle still attached.

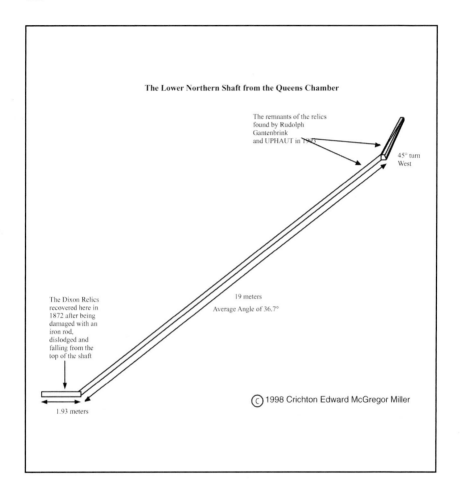

The Lower Northern Shaft from the Queens Chamber

The remnants of the relics found by Rudolph Gantenbrink and UPHAUT in 1993

45° turn West

The Dixon Relics recovered here in 1872 after being damaged with an iron rod, dislodged and falling from the top of the shaft

19 meters
Average Angle of 36.7°

1.93 meters

No further examination of the shafts in the Queens Chamber had been carried out until 1993, when Rudolf Gantenbrink developed a remarkable robot known as Upuaut 2. The robot was designed to explore the shafts and to install a ventilation system in the Pyramid of Khufu, to reduce humidity within the pyramid.

Robert Bauval suggested in *"The Orion Mystery"* that a priest might have placed the items on the other side of the shaft. However, it is obvious from the photographs taken at the top of the shaft by Rudolph Gantenbrink's robot, that they have fallen from the top and that there are remaining parts still lodged there. Therefore, it is my contention, in the face of overwhelming evidence, that the relics were placed in their position at the top of the shaft at the time of the construction of the pyramid.

During the writing of this book, it is reported that the piece of cedar rod has been located in Scotland by Robert Bauval and that he is currently trying to carbon date the find. It is unlikely that carbon dating will reveal an accurate age however. The problem is that the cigar box will in all likelihood have contaminated this relic. It remains for the current Director of Egyptian Antiquities, Dr Zahi Hawass, to allow the wooden slat at the top of the north shaft of the Queen's chamber to be recovered and tested. I believe that the archaeologists will be found to be correct in their current estimate of the date of construction and that it will be found to be in and around the start of the age of Aries, some 4500 years ago. The reason that I believe this to be so, is that the bronze hook is also a representation of the sign of the zodiac for Aries. This will be argued, of course, as current belief is that the zodiac as we know it was a Greek invention. I intend to show later on, evidence that this thought process is in error. Gantenbrink was subsequently to make the most accurate internal survey of the shafts to date, on behalf of the Egyptian Government with the approval of Dr Zahi Hawass, the Director of Antiquities of Giza. The operation was a success and a credit to Rudolf Gantenbrink. The discoveries made were remarkable and increased the understanding of the construction of the pyramid (for a detailed account of Gantenbrink's activities, see Part II of '*Secret Chamber*' by Robert Bauval). The robot is now housed in the British Museum after a personal presentation by Gantenbrink, (although it is only kept in storage and not on display), and the details of this survey are public knowledge.[xxxv] On the video footage of the shaft taken by Gantenbrink's Uphaut 2 in 1993, the remainder of a long wooden rod can be seen, from which the 13-centimeter piece had broken off. There can also be seen an object that appears to be a rectangle of wood or metal, with two corresponding holes to match the rivets on the metal instrument.

Rudolf Gantenbrink was unable to explore the shaft further for what have turned out to be political rather than technical reasons. Having collated the information available from the relics found in 1872 and those seen in the shaft in 1993, theoretical reconstruction of the instrument can be achieved.

The Relic items for reconstruction are as follows:

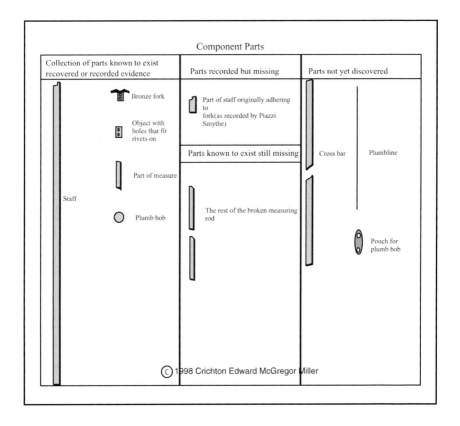

The objects found in 1872:

- Plumb bob.

- Fragment of scale rule (now missing but recorded by Piazzi Smyth in his diary entry).

- Bronze or copper fork with fixing rivets attached. Originally with part of the handle still adhering.

The objects found in 1993:

- A piece of wood with holes that match the rivets on the bronze item.

- A 2 meter plus length of wood resembling a staff with a section missing.

- Various pieces of unidentified material, located in two areas of the shaft.

- A large rectangular object can be seen at the upper end of the shaft attached to the 2-meter length of wood.

With this component list, it is possible to assemble the known parts in a logical format. Having completed the initial assembly and understanding the principles of the instrument earlier outlined, it can be established what components are missing which are needed to complete the instrument and turn it into a working model, proving my hypothesis.

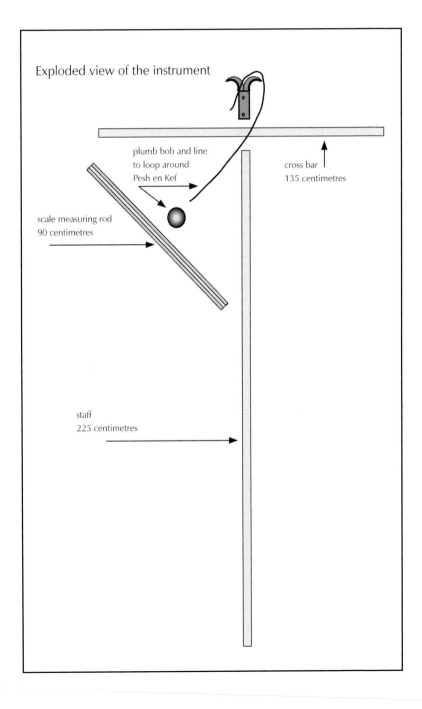

Exploded view of the instrument

plumb bob and line
to loop around
Pesh en Kef

cross bar
135 centimetres

scale measuring rod
90 centimetres

staff
225 centimetres

Assembly

1. The metal fork appears to have been designed to attach to one end of the long staff by a half housed joint held by the two rivets on one side of the hook.

2. The plumb line is looped over the fork, on the top of the staff, by way of a slipknot. The working end returning over the side of the fork opposite to the scale, this allows the line to cut the apex of the joint between the cross arm and the staff precisely.

3. A pouch or net is fixed to the opposite end of the plumb line, to hold the plumb bob.

4. A cross bar is mounted on the fork, at right angles to the staff and fixed with the remaining two rivets at the front of the instrument (I believe that this is the purpose of the object with two corresponding holes still in the shaft).

5. One end of a measuring rod is fixed at 45° to one arm of the cross bar.

6. The other end of the measuring rod is fixed at 45° to the upright of the staff.

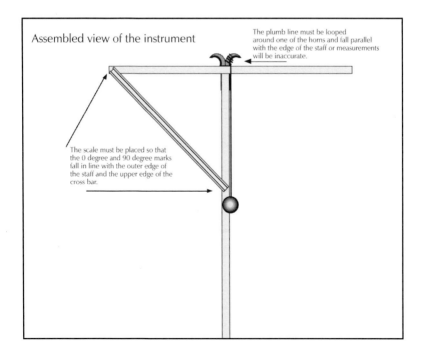

Assembled view of the instrument

The plumb line must be looped around one of the horns and fall parallel with the edge of the staff or measurements will be inaccurate.

The scale must be placed so that the 0 degree and 90 degree marks fall in line with the outer edge of the staff and the upper edge of the cross bar.

MEASURING ROD

It is my opinion that the Egyptians had sufficient knowledge through their established fractional and decimal mathematical system to allow construction of the measuring rod. The evidence of these mathematical abilities have already been proved and published by Sir William Flinders Petrie, when he surveyed the Pyramid of Khufu in the late eighteen hundreds. He also stated that the Royal Cubit measured 523.95 millimetres or 20.6 inches.[xxxvi] It has been surmised that the Fourth Dynasty builders divided the cubit into decimals.

Sir Flinders Petrie also named a unit of measurement used at Giza as a digit, which is constantly accurate to one tenth of a millimetre (see '*Temples and Tombs of Giza*' WMF Petrie). 1.75752 cubits = 9180 digits = 918 millimetres = 90 centimetres approximately. Sufficiently close enough for centimetres and millimetres to represent degrees and minutes for the purposes of this experiment.

EVIDENCE OF TOOLS

Sir Flinders Petrie also established that no tools were going to be discovered and described the reasons why.[xxxvii] The tools of the ancients were sacred and belonged to the Royal Family, losing one was deemed to be a serious crime and could have resulted in the loss of life for the unfortunate individual to blame. Therefore, the types of tools employed in the construction of the pyramids were only discovered through analysis of the marks on the stone. As a result, it was possible to deduce that there was metal in the tools used by the remnants left in the cuts. According to Sir Flinders Petrie, bronze and copper saws and drills of various designs were employed. Some with precious stones for tips and accordingly, would have been very valuable.

SECRECY

I believe that tools used in the construction and surveying of the pyramid, were owned by Master Craftsmen, who practised their trade within the membership of Guilds and that their tools and skills were precious to them. Whereas I do not think that the craftsmen were able to apply the knowledge of the full abilities of the cross to the construction of spiritual buildings, I am sure that they were aware of the application of such an instrument by the chief architect. These craftsmen were trained in the understanding of spiritual knowledge as well as the practical side of their individual trade. I believe that the same was true of the later generations of craftsmen employed during the great Cathedral building era of the Middle Ages in Europe. I further believe that this instrument is tangible proof of the ancient Egyptians' depth of knowledge in the disciplines of astronomy, surveying and navigation hitherto underestimated.

It is my opinion that this knowledge may have been kept secret for the following reason. The Trade Guild or Society members were sworn to secrecy during or after their apprenticeship, so as to protect their knowledge from the uninitiated. Later, this position of power and prestige was reversed and the religious authorities, from the destruction of the Library of Alexandria through to the seventeenth century AD, were to persecute those who practised anything of the ancient arts and crafts. The actions of the Spanish Inquisition in Europe and the destruction of the Aztec, Maya and Inca civilisations by the Conquistadors can ably demonstrate this. There is good reason to think that the Celtic cross was ancient before the builders of the pyramids used it, and it originated from a knowledge that was pre flood.

I think it was mainly mariners who survived the disaster that destroyed the ancient world. The evidence that this would be the case is self evident to modern mariners because it is the land that sinks ships more often than the sea. Most ships stay in deep water away from the shallows, because the waves build shorter and steeper as they approach the shore. An ocean roller breaks with great violence upon the beach, but when it is in mid ocean, it becomes a long swell that a ship can more easily cope with. The most likely peoples to survive such a cataclysmic event like the one that shook the planet ten thousand years ago, would have been those plying the oceans for trade.

It is those very people who, for the very basic reasons of survival, would have understood astronomy, mathematics and geometry. When they returned to shore and the ports that were once their usual places for docking and trading, they would have found nothing but mud and carnage floating upon the sea. The civilisation that they left would have disappeared beneath hundreds of feet of water never to be seen again. Only some land people may have survived and there are legends of this being the case, but the majority of all species and different kinds of mankind such as the Neathanderal had perished along with the giants that walked the Earth such as the mammoth, giant elk and sabertooth tiger.

Try to imagine what would survive our own society if such a disaster were to happen. Not immediately, but after a few hundred years there would be nothing left of our infrastructure or evidence of our knowledge to pass on to the children born afterwards. The image of our not too distant ancestors being ignorant cavemen is understandable in the eyes of the explorers who discover their remains. But let me point out to you, if a disaster like that occurred today, those of us who survived would also be reduced to living in caves and making crude implements to hunt and cook. What would those of us who had children after the cataclysm tell them of our past? We would describe in stories how our fore-fathers lived, we would talk to these little ones as they huddled around the campfire, of wonders that they would never see, of men who could fly, great wars, mighty ships and powers beyond belief. We would tell how we were on the brink of extending life and making hybrid creatures similar to the types in Greek mythology. All these myths are written in ancient manuscripts such as Sanskrit from India, the Bible, the Torah, the Koran and spoken all over the world in word and song to this day. We would say that we could talk to each other across the oceans and fly to the planets, and we would tell them that the sun was the source of life and that God destroyed the Earth because of man's wickedness. We would give them religion and frighten them into not changing the stories, but to pass them down generation by generation until our species had recovered enough to start to use the information that was in these myths, and climb back from the mud and slime of the abyss. Long after we had died and the generations had passed on through time, some of these poor souls would end up sacrificing their own to appease the powers of Nature, believing that if they made such sacrifices, such a horror might not happen again.

I think many may have lived a life on the oceans, for thousands of years, fearing to settle upon the land in case they were destroyed. Those who did live upon the land did not live in constructions of wood and stone, but travelled with flocks as nomads and hunter gatherers, knowing that they were safer living in that way. But all through this time of slow recovery for the last ten thousand years, they passed on knowledge of nature, astronomy and astrology through their initiates and apprentices.

A knowledge that was discovered in ancient times and in a world that we would not recognise today, before the flood, when the Earth was burdened with ice on most of the northern hemisphere and the seas were much lower than now. There is much argument and debate in this area of research, because of the similarity of types of arrowheads known as Clovis, which are to be found in Europe and the Americas dating to Palaeolithic times.*xxxviii* This has been the life work of a remarkable and learned academic who has given me freedom to quote his research. The following is an example of the remarkable investigations of Dr

Cyclone Covey, of Wake Forest University, Dept. of History, PO Box 7806, Winston-Salem, NC 27109.

THE SOLUTREAN CONNECTION

By Doctor Cyclone Covey[xxxix]

"Immediately on discovery of extinct-animal-associated Folsom, Clovis, and Sandia projectile points (in reverse order of their production) anthropologists recognised West-European resemblance and possible derivation, even across the Atlantic, but could hardly outgrow an assumption that all Paleo-Indians entered the New World afoot from Asia via a verifiable Bering land-bridge and imaginary ice-free corridor during the Wisconsin Glacier, nobody before Clovis points. Earlier Sandia points more exactly match Solutrean, regarded too early, Atlantic crossing ruled out. We here critically reconsider Solutrean impetus and feasibility and examine key sites that predate Clovis which indicate Atlantic transit quite early enough though not specifically Solutrean. Realisation by 1962 that Clovis points concentrated more in the east than in the west where first identified, coupled with recent discovery of pre-Clovis points in Virginia and South Carolina, has reopened debate about West-European relationship to America before 9500 BC This essay does not prove Solutrean colonisation but exposes the religiously held myths that prejudgmentally dismissed and desperately resist that scenario"

This is the kind of knowledge that is hidden away for us to find and it is hard to deny pre flood transatlantic crossings by ship in the face of this evidence. It is through seafaring that we began, and this is the knowledge that I believe we have received in fragments down the ages and built the foundations of our present civilisation upon. I also believe that the Palaeolithic peoples, followed by the Neolithic people, first used the cross. Other ancient civilisations inherited the cross, followed later by the Sumerians, Babylonians, Egyptians, Celts, Aztec, Incas, Mayans and lastly, in great secrecy, The Knights Templar to sail the seas and for the construction of their religious and astronomical buildings. I consider that the relics found by Dixon in the Queens Chamber were part of an originally complete instrument that had been sealed in the pyramid by the architects, in the same manner that masons and craftsmen made their mark on their stone masterpieces. This tradition is carried forward to this day and is usually found in foundation and corner stones of buildings in the form of Mason Marks. After the loss of this instrument again, our ancient ancestors resurrect it in our time so that we may wonder and learn at its use.

APPLICATIONS

(See appendix 1)

I intend to show you that in the able hands of a trained operator, this simple instrument can be used for the following purposes:

1. The Instrument can be used to take remarkably accurate astronomical measurements.

2. Measuring distances for chart and map making, including measuring the circumference of the Earth.

3. The Instrument can be used for Civil Engineering projects.

4. Timekeeping and the calendar.

5. Navigation.

But before we do this we must orient ourselves to the Egyptian and Biblical concepts of up and down.

CHAPTER 5

DISCOVERING TIME

WHERE IS UP?

The world is mostly made up of seawater and very little land. It is also true that in pre flood days; there was less sea, more land, a great deal of ice and cooler temperatures. The land based ice fields of the ice age, drew the water from the oceans and lowered the sea levels by at least three hundred meters from the current level that it lies at today. A clue to this knowledge is mentioned in the Old Testament of the Bible. It can be read as such:

> *"There was water above and water below and the firmament in between."*

Most people have difficulty in understanding this concept to such a degree that eminent researchers are looking for water in outer space. There may be water in outer space in all sorts of forms, but that is not what the ancients meant. They had a different concept of up and down, as we currently understand it. They saw the poles of the planet as up and down and not the spot over the head of the observer. If an observer is standing on the Earth at the latitude of 23.4° at the summer solstice, then the top of his head is pointing out into space in the direction of the ecliptic plane and not up in the conventional sense. We are stuck by gravity to the point in the curve of the sphere of the Earth at which we are located and like the cross, our feet point directly at the centre of the planet so as to maintain our balance. But up is not necessarily up in the sense that the ancients meant when they wrote their books of wisdom. You would have to stand at the pole to be directly up or down in that sense. It is said that they did not understand this and yet the Essenes wrote of the terrible ice in The Dead Sea Scrolls. The Essenes lived in one of the most arid climates and places on Earth where heat is constantly high and ice, let alone *"terrible ice"*, is unheard of.

Where did they see ice?

The flood occurred when the ice fields melted with sudden ferocity and violence. Do not think that what some academics state is true and that the flood was the result of a localised weather condition in the Mediterranean or the Nile. Use your imagination. The ice melted suddenly, according to modern science and raised the sea levels by hundreds of feet. *Hundreds of feet! Do you realise what that would mean if it happened today?* Remember the scenario that I related earlier on this subject. This is the same story as told in the bible, but in different terms, *"God caused the flood because of the wickedness of man"*. The same old story repeats itself in our time. We, the ordinary people, are blamed for global warming by our wickedness and excess consumption of energy. The scientists and politicians point this gross accusation at us alike and we are expected to take the blame again. These are the same scientists, corporations and politicians that made money and fame from inventing things in the first place and selling them to us. This knowledge does not absolve us of our duty to our planet, but is not our fault and we should not accept the guilt or burden our children with the responsibility of such nonsense.

Well, the sea levels are rising again as the ice melts, and it is not us who are causing it, it is the Sun. It is the sun, as it was the last time. The ancients saw the Sun as the Creator and scientifically, they were right. The entire planet and all that live and breathe exist only because of the sun, including you and I. If solar radiation increases, the ice on the poles will melt and the ozone layer will thin. More forest fires ensue and CO_2 emissions increase. The sun affects the centre of the Earth as well and its variations and storms affect the magma and magnetic field below the crust causing earthquakes and volcanoes. Volcanoes emit more CO_2 when they erupt than we could possibly emit with our puny attempts. If solar radiation decreases, then ice will form on the poles, the sea levels will drop as the water is taken up and general cooling will be the result. All these are cyclical effects and are part of nature and time, and the Mayan people knew this with their time keeping skills to such a point that they were able to predict sunspot activity and solar magnetic reversals. Consider their prediction that the Northern Hemisphere will be destroyed by fire in the year 2012.

2012 is the next predicted height of solar activity. The ancients knew that the Earth and all that is on it owes its physical existence to the rays of the sun. The sun is the material creator of all physical manifestations of vegetable, animal and mineral. It is and always will be the creator of life in material form and it governs time upon the planet on which we live.

It is my belief that all calendars originated through agriculture, mostly due to the observation that there was a time for planting and reaping that fell in line with the seasonal inundation of the Nile. This inundation is also caused by the sun and its affects on the weather systems that cause rainfall in the heart of Africa. We know that the Ancient Egyptians were able to keep time as can be demonstrated through the legend of Horus and his followers as detailed previously. Horus was the keeper of time and Horology, the science of time keeping as it is known today, originated from his name. There are many inscriptions and paintings of Horus recording time in writing. In modern times, the back of the American dollar bill shows the eye of Horus on a pyramid. The association of this is one of time keeping through the great pyramid and whoever designed this currency had knowledge of what I am about to reveal to you.

An ancient or modern observer using the cross and plumb line can find the ecliptic pole at any time of the year, in daylight or at night because the ecliptic pole is at 90° to the ecliptic. The ecliptic pole changes position only 2.5° over 40,000 years and therefore can be considered constant. The north ecliptic pole is always in the middle of Draconis (the Dragon or Serpent) and can be found between the stars Noedus 1 and Noedus 11 in a direct line from Polaris to Eltanin. Accordingly then, we are still influenced by the Serpent as we were in ancient times.

In the depiction of the Celtic cross above is one of the most important signs for the serpent. It is reproduced all over the world and can be clearly seen in Mayan Glyphs as well. Around the circle is the interweaving serpent that eats its own tail and is the ancients' description for the constellation of Draconis.

This is essential knowledge for those seeking the way in which the ancients thought. For those who do not understand this yet, let me take this opportunity to familiarise you with a little astronomy. Precession causes a 46.8° wobble of the Earth over 25,960 years, this discovery was credited to Hipparchus and later on to Sir Isaac Newton. The effect causes the area of the sky over the poles to change with time, at present our north polar star is Polaris. 12,500 years ago in the half cycle of precession, our polar star was Vega. The precessional wobble is accented or decreased every 18 years by the effects of nutation which is the effect of the moon's gravitational field on the Earth causing a small wobble and can spoil the calculations of precession for the observer. Due to the obliquity (tilt) of the Earth's angle against the ecliptic, as caused by precession, the north ecliptic pole is presently at an angle of 23.4° to the celestial or North Pole. If you deduct the 23.4° from the 90° between the equator and the North Pole you will come to a startling result, you will find that the ecliptic pole rotates around the Earth directly above the north latitude of 66.6°.

The following illustrations that I have made are designed to clarify this for you.

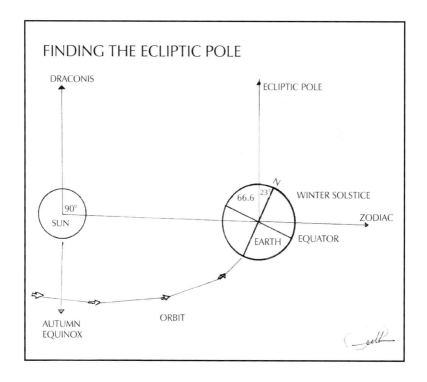

This conclusion is an amazing coincidence with the hidden astronomical and astrological numbers hidden in the Bible. Could this be the hidden message regarding the beast having a given number of 666? The ancient Egyptians also called the Earth "GEB" (the Beast). We know the Earth through a long and relatively gentle period as a nurturer and mother. This is so but the ancients were closer to the effects of the flood, when the ice caps suddenly melted and destroyed most of the flora and fauna of the planet, including wiping out human civilisation as it was then known.

The date given for this terrible event was 12000 years ago by the Dead Sea Scrolls, which were lost in the dust for over 2,000 years. They were found in the last century and the ancient knowledge kept by the Essenes and protected from destruction of Alexandria, at last, co-incides with the latest findings of modern science. How can you imagine such an event as the flood with its earthquakes, violent seas, tidal waves and the rain, which would have caused confusion, destruction and horror so terrible as to drive the survivors to the point of madness? There is geological evidence of tidal waves of such magnitude that they forced trees and animals into cracks high up cliff faces. If the ancient recorded myths are correct and the Earth suddenly tilted, from an ancient observer's point of view who might have been watching the stars, the sky probably did roll up like a scroll. As I already said, the constellation Draco or the dragon is where the heart of the ecliptic pole is located and apart from a few degrees of shift over 40,000 years, this location is virtually unchanged. The celestial pole is therefore an invaluable tool for the ancient navigator or astronomer.

These skilled observers would aim the sighting arm of the cross toward the sun at its noon azimuth and always know that the upright arm at the top of the cross would be pointing to the heart of the ecliptic pole.

THE ECLIPTIC POLE IN RELATION TO THE CELESTIAL POLE AND THE EQUATOR

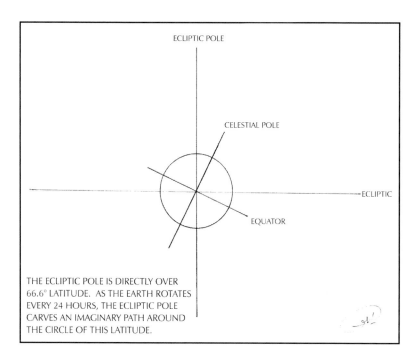

ECLIPTIC POLE

CELESTIAL POLE

ECLIPTIC

EQUATOR

THE ECLIPTIC POLE IS DIRECTLY OVER
66.6° LATITUDE. AS THE EARTH ROTATES
EVERY 24 HOURS, THE ECLIPTIC POLE
CARVES AN IMAGINARY PATH AROUND
THE CIRCLE OF THIS LATITUDE.

What I am about to reveal may be the answer to many questions. The north shaft of the great pyramid is at an angle of 26°. You must take into account that the latitude is 30° at Giza. This angle had me puzzled for many years, until I realised that the stars that could be observed by looking through the shaft from inside. Such pit observational methods were employed in Mesopotamia. The stars observed, however, would be 4° below the North Pole star and on the daylight side of the Earth. An observer, at the time of the building of the Great Pyramid, sitting deep within the north shaft, could watch the stars as they wheeled by and be rewarded, while observing the night sky through the small aperture, with a remarkable event. Once a night, the star associated with the weighing of the heart ceremony, Thuban, would appear for a moment and then pass out of view. These are the areas of the "Imperishable Stars" and they could be seen all the year round, never setting on the northern horizon. A perfect clock of rotating circumpolar stars whose angle in relation to the North Pole and the ecliptic could always be observed at night when the sky was clear. It is at this 30° latitude, that a slope of 52° becomes crucial in the following exercise. Once a year, the southern slope of the Great Pyramid of Khufu points directly toward the ecliptic pole.

Let me explain why to "those who have ears and eyes". I shall explain it by using the cross that built the pyramid in the first place; this lost architect's, navigator's, astrologer's and alchemist's fabulously simple instrument. The cross is the precious sextant, astrolabe and theodolite that pointed the way out to the cosmos for the souls of those wishing to gain heaven and return to the stars. Try to imagine a different sky all those years ago. The winter solstice was not the same as we see now with Orion high in the night sky. At the winter solstice at the end of the age of Taurus, the Royal star, Regulus in the constellation of Leo was seen at its azimuth in the southern sky, 6.6° and 396 miles south of Giza. Almost directly overhead the modern Aswan Dam and on the prime meridian. At midnight, it stretched away to the east for 1,800 miles, a Great Sphinx in the heavens for all to see. The great constellation of Leo rose at sunset on the eastern horizon and disappeared slowly over the western horizon at dawn. As this ageless constellation of time rose in the east, it would face the man made sphinx that greeted its arrival as the velvet blanket of night fell and the sun faded in its golden glory in the western horizon. The time could be measured by its angle in the sky as it passed from horizon to horizon, just like the hour hand on the face of a modern clock.

Regulus, not Orion, was the winter marker and it was measured with a cross and plumbline until the knowledge became lost again with the diffusion of the Archetype. Because of the effects of precession, the seasons move slowly forward over the aeons and the constellations change position in the night sky. "*Winter has become summer and summer has become winter, all is confusion*". This is the sad and woeful lament of those who have lost this vital understanding and depth of knowledge.

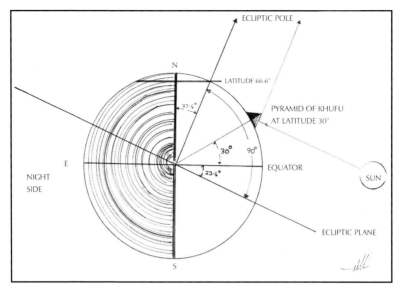

During winter solstice the sun is reflected directly at the ecliptic pole which is a latitude of 66.6°.
This is due to the location of the pyramid and the approximate angle of the slope of 52°.

So, as you can see from the diagram of the Earth on the winter solstice day at noon, the south face of the pyramid will face the sun and the southern slope will point to the ecliptic pole. The ecliptic pole will be precisely on the ancient meridian of 0° longitude at 66.6° north latitude, on the same side of the Earth as the sun and in nearly perfect alignment.

The Winter Solstice will have arrived 4500 years ago as the great clock of the pyramid prepares to welcome in the start of the Age of Aries and when Amen will become the living God. And what of the man whose number is that of the beast, could that have been Khufu? The pyramid of Khufu was called "Khufu's Horizon" as an interpretation of the "Time of Khufu." It was his time in the light of the sun under the Royal Star, where the Sphinx dominated the winter night sky. In all his mighty power as keeper of the time and place in the world, before joining the underworld, he either built or prepared a pyramid that also indicates latitude of the Serpent and the number 666.

So where is the connection with the Eye of Horus on the dollar bill? The faces of the pyramid were cased in polished marble and were reflective and designed to show an image of the sun when it was due south on the prime meridian. No one may look upon the face of "*God the Creator*", as they will be blinded, but to look upon the sun in a reflection is safer. Because the sun is lower in the winter than in the summer, at noon, it could be seen as a reflection low upon the southern face of the great pyramid. "*The Mighty Eye of Horus*" would be seen as a fiery image for all to see. As the year went by, the sun would rise daily up the face of this great instrument until it could be seen at the apex of the pyramid, on the Ben Ben stone at noon on the Summer Solstice as "*The all seeing Eye of God*". The Serpent would be in the night sky on the other side of the world. In all its glory, the sun would be seen reflected on this monument and its image bounced from the satellite pyramids and the others and between each other in the Giza complex at all times of the day and in all directions. The angle of 52° is precisely the same as that angle required of the reflection of the sun in a water droplet as it breaks the light down into its component parts and forms the glory of a second bow rainbow. Each face of the pyramid would reflect the sun as it rose in the dawn and set in the west at dusk. Causing an incredible light show of such proportions and a celebration of nature never to be seen again in the history of our species.

SOLAR OBSERVATORIES

The most basic instrument remaining from ancient times and passed down through the ages, is the sundial. It is an instrument extremely accurate for determining time and it is only in recent times that the railways in the United Kingdom stopped using it to determine local time. They developed the system of using sundials, because they were aware that on rail journeys across the countryside of England from East to West or from West to east, that there would be a variation in local time. To examine this, let's observe a journey from London to Cardiff, in Wales. This distance is at least 120 nautical miles, which is equal to 2° of arc. Since the planet is spinning from west to east, it follows that Cardiff is a measure of time behind London. There is a difference of approximately 2 minutes between the time at London and that of Cardiff. To show this practically, imagine that a signalman telegraphed the departure of a train from London at 12.00 Noon Greenwich meantime on the prime meridian, which passes through London, and it is Midsummer Day. The sun is over London and at its highest point in the sky directly south. Since the telegraph travels at the speed of light and the signalman relates the message that the train left at 12.00 noon, the receiver of the message would receive the information almost immediately. It is not noon at Cardiff, it is two minutes before noon and the sun has another two degrees to go before reaching its zenith. If the train is estimated to travel at

60 miles per hour and the journey takes two hours, then the train will arrive two minutes early if local time is not taken into account. An accurate method of obtaining local time in relation to the prime meridian time is to mark the sun's position at the time that the signal is received on the telegraph. Two minutes discrepancy could cause a substantial accident and subsequent loss of life, so keeping local time was important. I am revealing with this idea, the method of finding longitude using heavenly bodies. The question that then arises is what method could the ancients have used to signal the arrival of a heavenly body at the prime meridian?

IT'S ALL DONE WITH MIRRORS

They would have found this to be a much more laborious method than we do today, but then they were more motivated than we are, after all they built the pyramids. The answer is again simple, it was done with mirrors and beacons.

A modern beacon with a grate designed to hold fire on the Scottish borders.

Staging posts manned by watchers would have been initially set up using high points that were located at a measured distance from each other, and many of these are still to be found in the British Isles. On ordinance survey maps all over the country, there are hills named as 'Beacon Hill'. A strategically placed observer can signal to the next hill at the speed of light with a mirror in daylight and a fire at night. The distance from horizon to horizon is the only cause of any restriction that exists; after all, the Elizabethan English warned the whole country of the arrival of the Spanish Armada by using this method of signalling each other. Certain Amerindian tribes used fire and smoke signals to great effect, much to the distress of the settlers and their protectors.

The obvious answer is that when the sun reached its noon point over the meridian of the observer, that observer would reflect its light as a signal that it was noon at their location to the next observer, who would signal it on down the line. The same event could be recorded at night with the use of beacons and fire, when a star crossed the meridian. What we are establishing here is the concept of a clever but simple network of communications about time. This system is one where the ancient observers used the very sun as its own efficient signal-man, but to do that properly required the perfection of the instrument known as the sundial. To continue with sundials, since that is the essential ingredient of first finding local time in a simple way at any location. It is my understanding that in ancient Egypt, they used a form of sundial produced initially as a pole or stone obelisk. The Gnomon or vertical stake was placed in order to cast a shadow as the sun moves across the sky, and is considered to have been the first astronomical instrument, and was used already by pre-literate societies. With this instrument you can measure the altitude of celestial objects i.e., their angle above the horizon. Greek astronomers like Ptolemy used the Gnomon to determine the length of the year with considerable accuracy for the times. From the measurements of the sun's altitude and knowledge of the sun's position in the sky on any day in question, you can compute your geographical latitude i.e., the angular distance from the Earth's equator to your zenith, with the apex of the angle being the centre of the Earth. It would also have been observed that in certain positions, the shadow of the gnomon coincided with the annual Nile inundation and harvest time. In the light of this knowledge, seasonal predictability became possible, as did orientation to the cardinal points.

I believe that the Pyramids at Giza, as well as others throughout the world, were and still are perfect sundials. It is my opinion that any observer, who is trained in this art of measuring the position of the sun, may easily tell not only the hour of the day, but also the month of the year from these ancient instruments. It must be remembered, however, that the cross and plumbline will also carry out this task and that the cross is the original archetype from which the Pyramids were made. Orientation to the cardinal points is extremely important in telling local time as local noon is measured from the moment the sun passes the meridian of the observer.

Many people are under the misapprehension that the sun may not be measured accurately because of the width of its disc. This error in thinking has been created by the use of optics in a sextant, where the image of the sun has to be brought down to the horizon so as to measure its centre. This is not so in the case of a shadow, which is focused precisely providing you are aligned with the ninety degree angles of North, South, East and West and to be sure of that, you need to understand solstices and equinoxes.

THE SOLSTICES AND EQUINOXES

By marking a rising point of the sun in the east, using a natural landmark on the horizon for alignment, it could be seen that the sun would rise predictably and progressively at points north and south along the horizon. When the sun rose at its most northerly point along the horizon, the sun was at its zenith, the temperature of the season at its highest and it was the longest day of the year. This day that happens once a year became known as the Midsummer solstice. Conversely, when the sun rose at its most southerly point on the eastern horizon, it was at its coolest and the day was the shortest of the year and this is termed the midwinter solstice. The middle trajectory of the sun between these two points, spring and autumn, would eventually be known as equinoxes. Equinoxes mean equal and equal means that the sun and the zodiac ecliptic constellations travel at the same altitude in the sky in relation to the observer. The effect of an equinoctial event is quite dramatic because the Earth has become side on to the sun as it travels round the path prescribed. The signal of this event is a sudden change in the weather conditions and for a sailor it is the time of storms and rain. The observation that is perceived that this event is taking place is quite obvious in the night sky. The sun rises higher in the east and sets lower in the west and the opposite is true in autumn. These events are responsible for our changing seasons purely because the Earth is at an angle to the sun, known as the angle of obliquity. This oblique angle is caused by precession of the equinoxes and should this system change for any reason, then this would be a very different planet indeed.

A priest predicted these important times of the year and would ultimately have the responsibility of indicating the times for planting and reaping, accompanied by the associated feasts, celebrations and thanksgiving. In the Northern Hemisphere, we are used to the idea of four distinct seasons, but in Ancient Egypt, there were three. Predictions were made by the position of the stars and the sun in relation to markers such as stone or wood circles or landmarks such as mountains in the surrounding area. This is why most henges in the Northern Hemisphere are on flat plains surrounded by hills or on hills with a clear view of the horizon. These henges which display a high degree of sophistication are impossible to set up without the aid of the cross and plumbline. Evidence of these observatories and clocks may be seen in continents and countries all over the world and on both sides of the equator. That they were all built in prehistoric times is self-evident.

The following examples of ancient measuring earth works and henges have been chosen to highlight this skill and dedication to time keeping, which was displayed by our forefathers.

CIRCLES AND STONE CROSSES

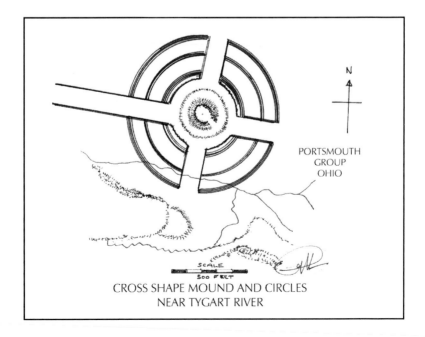

PORTSMOUTH
GROUP
OHIO

N

SCALE
500 FEET

CROSS SHAPE MOUND AND CIRCLES
NEAR TYGART RIVER

By using the knowledge gained from the ancients who used the cross, logical progression led to the erection of stone and wooden circles where the same predictions could be made in a manmade environment instead of a natural one that may have otherwise proved difficult in the local environment. The remnants of these objects can still be seen in many parts of the world. The remains of one can be found in Portsmouth, Ohio, America, which is laid out as a representation of a Celtic cross (as can be seen in the first diagram). Another example is in Pickaway County, Ohio. These workings predate the current opinion of the construction epoch of The Great Pyramid of Khufu.

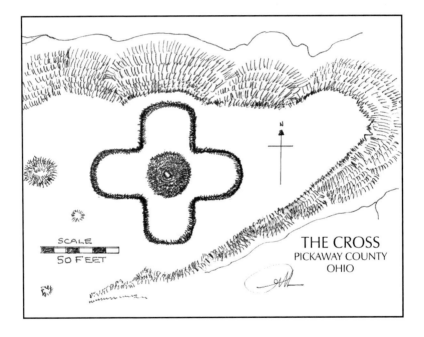

The most important circle illustrating the existence of the theory of the cross as an archetype is on the Isle of Lewis in the Outer Hebrides in Scotland,[xl] known as Callanish. The stone circle is built in the shape of a Celtic cross and the legends tell of it being built by a priest king dressed in a feathered robe and accompanied by a fleet of ships with black men as crew. It is officially dated to the early Bronze Age, about 5,500 years ago. Therefore, it would appear that it is a Celtic cross, which was constructed earlier than the Giza Pyramids. Does this construction indicate an ancient, multi ethnic group of navigators with a thorough astronomical knowledge?

Recently, researcher and author, Simon Cox visited the site of Callanish, these are his thoughts:

My one overriding feeling after having been to the remote corner of these isles that is Callanish, is one of wonder and awe. The stones there stand testament to the achievements of an ancient civilisation more in touch with their surroundings than we can ever claim to be. Over twenty stone sites of one form or other and another twenty-cairn sites dot the landscape in what must be a true representation of the Roman's Hyperborea. The feeling of total awe at this achievement is so strong that I was left speechless at many of the sites we visited. Margaret Curtis, the 'Omm Sety of the stones', as I have dubbed her, was on hand to relay her thirty years of experience at the site. She pointed out the lunar significance of the main circle and also the area known as the 'sleeping beauty', a recumbent goddess figure in outline dominating the horizon. Callanish is truly an amazing place and one at which it became apparent to me and my companions, that some form of sophisticated surveying technique must have been utilised, instruments for which had to have been used. At this site, more than any other, I felt that the idea of a device such as Crichton's cross was tangible. Surely they had to have used such a device in the layout of this place.

We left Callanish reluctantly, vowing to return soon. I have undertaken a pledge to Margaret and Ron Curtis to help them with the publication of their mountainous research data and material so that a wider audience can sample the awe and wonder of this special place.

WATER OBSERVATORIES

A further innovation of the circle system was employed when water observatories were developed which allowed the observer to predict the time that a heavenly body cast its reflection against marks set in the right position around a pool. Again proving that types of measuring circles were part of life in ancient times. This form of clock can be seen in the recently discovered wooden "Sea Henge" on the East Anglian coast of England at Holme-Next-The-sea[xli] which constitutes a ring of 55 wooden posts with a central upturned oak root. This Henge is thought to date back to over 4,000 years ago (see photograph). A good example of the method using a large cross shaped pool, would be the water features in front of The Taj Mahal in India. In this case the axis is aligned North to South, and Cyprus trees are planted along the sides.*xlii* By watching the reflection of the sun or moon as either body makes their way from east to west across the pool, the progress of the reflection of the heavenly body could be measured against objects around or along the sides of the pool, thus predicting the time down to minutes, or fractions of minutes.

Sea Henge on the North coast of Norfolk, England.

The breakthrough in this system was that the moon could be employed for night timekeeping on several days of the month. The drawback being that wind rippling the surface of the pool would obscure accurate vision. Nevertheless it added a further dimension to the overall system. The potential to tell the time when night fell.

DEVELOPMENT OF A STAR CLOCK
(A vital ingredient for Navigation)

The problem at night is that unlike daylight hours, when a shadow is formed by the sun, there are no shadows produced by the stars and planets. The challenge would have been to devise a method of predicting time by observing the stars over a period time. The ancient observers will also have noticed that the stars on the ecliptic follow the same rules as the sun, except that their position on the point of rising from the eastern horizon is opposite to that of the sun at solstices and the same as the sun at equinoxes. I have found that the seven-step pyramid may be used in the same way. Six, seven and eight step pyramids may be found in Egypt and South America, with several small ones to be seen in the Canary Islands. The pyramids in the Canary Islands are aligned east and West in the discipline of this ancient knowledge.

The island of Las Palmas (one of the Canary Islands), is located furthest out from the African continent; it lies in the direction of America and has a pyramid on its West Coast. This westerly observation post would be ideal as an embarkation point for voyagers heading out towards the West Indies. Having found local time against the prime meridian at Giza, a mariner would be able to commence the transatlantic voyage with confidence.

The pyramid that I visited on the East Coast of Tenerife in 1999 was a combination unit allowing visual sightings of stars and planetary bodies at night, whilst permitting accurate angular measurement of the reflections of the sun on the ocean during the day and the moon at night. In my opinion it is a version of a water clock observatory. Its use would also be to find local time through astronomical observations to aid pre-historic mariners attempting a transatlantic voyage. I found it interesting that at the museum on the island, which is right beside the pyramid, is the location of Thor Heyerdahl's papyrus boat Ra that he used to show the capabilities of the Ancient Egyptians to carry out a transatlantic expedition. Ra is displayed along with a model of Christopher Columbus' ship The Santa Marie. Do the curators of the museum know the secret or are they just driven by associations? They are insistent that the visitor makes up his or her own mind over the displays showing Neolithic spirals in stone, similar to those found in Scotland and America. I shall cover the use of local and prime meridian time and its importance to mariners in the section on longitude. However, to be able to lay out and build the pyramid with such accuracy in the first place, would without doubt, require the use of the cross and plumbline which would be the only type of instrument capable of obtaining accurate vertical angles.

THE SEVEN STEP PYRAMID CLOCK

This remarkable invention enables the user to be able to tell the time at any part of the day or night providing that the sky is clear. It is designed to illustrate the rise and fall of the planetary and stellar bodies throughout the year as the ecliptic rises and falls dependent upon the season. By standing at a pre-designated position, the observer can determine not only the month, but also the week and the day. This is achieved by predicting which of the seven steps of the stone pyramid the observed body will touch or be eclipsed by. In my opinion, the Step Pyramid of Djoser is a good example of this type[xliii] and was reputedly designed by the architect Imhotep.[xliv] Remember though, that most of the pyramids in Mexico and the Yucatan are also step pyramids and yet the present thought pattern of academia is that this is a coincidence.

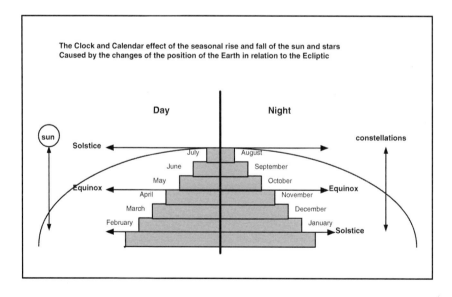

With the positioning of shadows, the pyramid can be used as a sundial and the shadow cast is incredibly accurate at this size for measuring units of time by day. At night, an observer can tell the time with an accuracy of seconds, as a pre-dicted star or planetary body on the ecliptic or celestial equator touches the edge of the pyramid. It must be remembered that at the equinoxes, the sun and the stars are at an angle because of the obliquity of the Earth, as any one who has watched the sky at this time will have observed. In order for the constellations to be easily identifiable, the stars were grouped together in universally understood shapes and named accordingly as signs of the zodiac. Now we start to see simulacra used in identifying the stars in the same way as we earlier saw in Maureen Palmer's article of simulacra spirit forms in rock. This caused them to consider that the position of the stars in heaven were also reflected upon the Earth originating the "*as on Earth so below*" statement that is so well known.

Here are the examples of forms that they used for identification. The constellation of Leo looks like a Lion and the constellation of Taurus looks like a bull. Orion the Hunter, looks like a man and was the great marker of the constellations of the ecliptic. Orion is accompanied by his hunting dog Sirius, which is also shown in the medieval world map known as the Mappa Mundi, which is kept in Hereford Cathedral, England. Orion is the master of the twelve constellations of the ecliptic zodiac, which may be seen travelling from east to west and disappearing over the western horizon every night. Above thirty degrees of altitude, in the North Polar Region of the sky, are the constellations known by the Ancient Egyptians as "The Imperishable ones". They are known as such because they never set over the western horizon and are constantly on view at night. They are recognisable by form as well and the Egyptians named them on such designs as the Zodiac of Dendrah. An example of this northern type of constellation would be the well known Great Bear, Plough or Big Dipper, which we use today to sight the current pole star. It is interesting to note, that by following the pointer stars in reverse and away from Polaris, would lead the observer to the star Regulus in the ecliptic constellation of Leo. If you try this in winter, you will be able to tell where Leo is after it has set over the western horizon. If you know what constellation of the ecliptic the sun is in, you will be able to tell its position on the other side of the world and predict the time of the dawn from your position. These are all signs of the zodiac and were essential for astronomy, astrology and telling the time.

CHAPTER 6

GODS IN THE SKY

THE ZODIAC

The zodiac or astrological chart is believed to have been invented by the ancient Sumerians, in the form of a star clock, as early as 4000 BC.[xlv] Each sign of the zodiac was calculated to be exactly 30° between the end of one sign and the beginning of the next and is aligned along the ecliptic. The twelve constellations of the zodiac create the 360° circle, and were named after various animals and symbols in the same form as simulacra, so an observer could identify them easily. Today, the zodiac is still used in astronomy and astrology although the general population associates its use with fortune telling.

Fortune telling using the zodiac is an ancient custom and generally considered to have been the skill of Gypsies or *"Egyptians"*. The art is practised by first establishing which star sign the client was born under and this star sign is that sign which the sun was in on the day of their birth. The signs are divided into four types according to their group, the four types are Earth, wind, fire and water. It is believed that each person is affected by the quality of radiation received from the sun at the time of their birth and furthermore it has a ring of truth since the sun itself is divided into four parts of negative and positive magnetic component. There are in fact thirteen signs of the zodiac and the thirteenth is Orion, which lies between Aries and Taurus.

This is part of the meaning of the story that is related in the bible about Jesus and also in the legends of King Arthur and the round table. Both of these leaders have twelve disciples or twelve knights and with Jesus or King Arthur as the leader, each group numbers thirteen in total. Jesus and his disciples and King Arthur and his Knights are designed to bring the hidden concept of the Zodiac and Orion to the attention of the reader. Orion is a very powerful image and has been depicted in many ways. Samson and Delilah are one such story, which tells of the temple being brought down after Delilah cut Samson's hair. This has often been translated as meaning that there was a change during the age of Leo when Venus passed over the head of Orion. The date is considered to be in the age of Leo because Samson wore Lion Skins. A similar story is related with the beheading of John the Baptist and that time was considered to be when Venus passed across the face of Orion. The date here is considered to be in Aries because John the Baptist wore rough sheep's wool clothing. It is obvious that a great deal was written in the bible with hidden meanings pointing a finger at astrology and time.

The Egyptian priests dressed in a kilt to represent the image of this sign and it is often thought that the Scottish kilt is a derivative of this. Orion is recognisable by the belt at the top of the kilt and the three stars below this kilt are often taken as a symbol of a large phallus. Orion was used as the main marker at midnight when seen due south and the rest of the signs followed as the year progressed until returning to Orion. The length of a month was measured by the daily progression of the sun across the face of a particular constellation or sign until it entered the next one. This system was also used in conjunction with lunar observations. Viewed on the 20th December in our present age in the United Kingdom, the point when the star Al Nitak on Orion's Belt breaks the eastern horizon is at 18.00 hours. The constellation of Orion will be at its zenith due south at midnight and will set in the western horizon at 06.00 hours on the 15th December 2001. Time could be measured by the predicted rise of a particular star in the eastern horizon, by its azimuth position due south, and by its setting in the west. More accurately though, time could be calculated by the number of degrees measured between these points, but only on the supposition that they had an instrument enabling them to measure angles.

This can only be achieved with an instrument like the cross and plumbline that could measure the precise angle of that star, as it travelled across the night sky. A hand held instrument of the type found in the Queen's Chamber (see the Dixon Relics) could measure within 3 minutes of arc. A larger instrument coupled with a tripod and a scale 5 meters long would provide accurate measurements to within 30 seconds of arc.

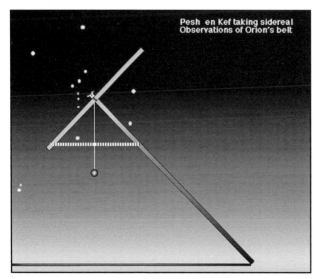

Pesh en Kef taking sidereal Observations of Orion's belt

The cross of the House of Ram taking observations of Orion as it rises in the East.

Notice the observation notch created by the bronze sign of Aries or Amen at the top of the cross. This was made in bronze so that the reflection of firelight would make it visible against the night sky.

In order for the ancients to be able to understand the length of the year, they would also have had to understand that the star moved 1° westward every twenty-four hours, as the sun does when it traverses the monthly zodiac sign. The Ancient Egyptians calculated the year as being 360 days plus 5 festival days and according to Egyptian Mythology, the God Thoth (the god of wisdom) won the extra five days from Re (sun) on behalf of Nut (moon or sky goddess).

> *Re had forbidden Nut to conceive children from her brother Geb (earth) on any of the 360 days in the old year. Thoth won the extra five days from Re in a game similar to chess. So Nut was able to have children, five in all, each conceived on the five days outside the old year of three hundred and sixty days. Her children were Osiris, Isis, Seth, Nephtys and Horus.*

This could not have been established as an astronomical, astrological and horological story of how the year works, without a system of measuring like the cross together with a method of recording the observations. This question of how they actually measured angles has not been raised before as there has not been an answer, there is only the acceptance of the mysteries that have been left by the Ancient Egyptians. They are known to have measured the moon's cycles. They measured and named the visible planets. The word that springs to mind for modern day historians is 'how'? How was this measured so accurately? What instrument was available to them in order to be able to achieve this? I give them this instrument known as the cross as an answer and I am sure there is no other.

Although astrology has long been separated from the science of astronomy and academics are suspicious of it, astrology was far more fundamentally important to the ancients and consideration must be given to the subject. The astrological charts as we now know them were, and still are, used to identify birth types and other significant events and took into consideration the planets that were in the sign at that specific time. Again, the question has to be 'How' did they do this, how could they identify the star sign of a birth date? We know that any method used to calculate the movement of the stars would be impossible to invent without measuring equipment but also they must have known that the Earth was a sphere and that the Earth went round the sun. Let me explain why by first asking the question, "How could the ancients locate the position of a star sign in daylight"?

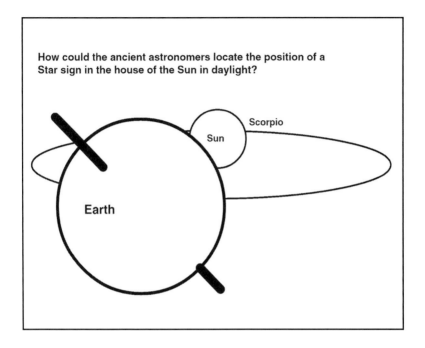

How could the ancient astronomers locate the position of a Star sign in the house of the Sun in daylight?

A birthday or event is said to happen when the sun is in the house of that sign. In other words, the constellation was on the other side of the sun and obscured from view. You cannot see the star sign of your birthday because it is daylight when you look for it and the sun would blind your eyes, but that night if you looked in the night sky, you would see your star sign's opposite number. If you look at the opposite sign at midnight and follow the sky north with your eyes toward the pole, that is where the sun is on the other side of the world. In the example above you will see Scorpio as the sign behind the sun during the day and that following night if you looked south at the ecliptic, you would see the star sign of Taurus. In these current times, a Scorpio sign is to be seen as a birth date in October. So the owner of this sign was born when the sun was in the House of Scorpio and which day would be applied to this would depend on how many degrees the sun was along the thirty degrees or days of that sign. At night, on that date, Orion would be seen on its way to culmination at the winter solstice and Taurus would be high in the sky at midnight.

I was born on the 25th August 1949 and I am a Virgo, but if I were to give the real date as the ancients knew it before it was mucked up by the invention of the Julian Calendar, my birthday would be the 2nd day of Virgo. I firmly believe that the ancients did not suffer from what is a kind of slide rule imposed by the calendar makers in semi-modern times and that they originally called the months by their zodiac sign. The result would be no confusion as precession moved the months backward as the zodiac moved forwards. The 27th of Scorpio would always be the 27th of Scorpio regardless of the climatic and seasonal changes that ensued because of the steady creep of the precession. Remember, for several months, your birth sign and the adjacent signs can not be seen because they are in daylight. Yet observers were able to declare what constellations and planets were between the Earth and the sun at any given time. This inevitably, requires sophisticated observations and recording methods.

Stonehenge was such a measuring system and if looked at in this way, using a cross and plumbline, plus recording where the planets and constellations were when out of view in daylight, great accuracy in prediction could be achieved. Modern astrologers, in general, do not make their own observations, but take the knowledge second hand from published almanacs. The ancient Egyptians did not only use the constellations of the zodiac on the ecliptic, but also used the imperishable stars such as the ones shown on the Zodiac of Dendrah of the Ptolemaic era.[xlvi]

The Zodiac of Dendrah

The imperishable stars do not set and at the latitude of Giza, a full 30° observation of these stars is possible during the night hours. More importantly, the position all of the ecliptic zodiac constellations that are obscured by the sun during daylight can be located using lines drawn from the pole through the imperishable stars to the hidden stars on the ecliptic. In other words, you can see the stars that are on the daylight side of the Earth when you are on the night side, by looking over the North Pole of the planet axis. An example of this is by following the pointer stars in the opposite direction from modern Polaris and you will come upon the Star sign of Leo. If it is in August, then the sun is in the house of Leo and that means you can work out the position of the sun when it is on the other side of the planet. If you use the cross and plumbline as an aid, you can measure the angle involved to minutes of arc. I know that an observer with the type of instrument that I have discovered can measure the declination of the stars to a remarkable accuracy and coupled with a thorough knowledge of the constellations on the ecliptic, tell the time to within three minutes at night.

The current argument against this concept of the zodiac signs amongst some of the academic fraternity, is that it was the Greeks who invented the Zodiac in its current format although a few consider that it was invented in Mesopotamia. However, there is evidence that this format is in fact pre flood and at least 15,000 years old. This possible evidence comes from some of the cave paintings found in France. Although it has been suggested that the age is wrong and that the drawings could be much older because of the errors recently found in Carbon-14 dating, which was caused by a variation in the amount of CO_2 in the atmosphere at that time. There is a cave painting of a bull facing right to left as Taurus does in the sky. It has only just recently been noticed that dots around the bull represent moon phases and furthermore there is evidence that above the bull's shoulder there is a cluster of dots that represent a star group. A star group that is still important to the Mayans who believe that their ancestors originated from this constellation. That group of stars is known as the Pleaides.

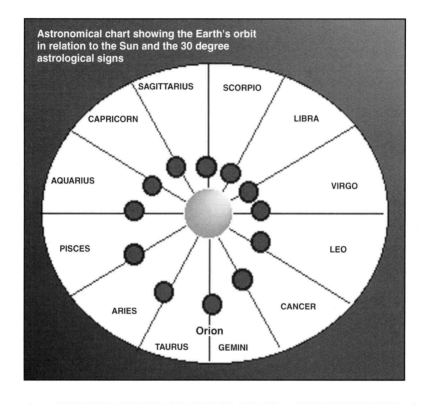

Astronomical chart showing the Earth's orbit in relation to the Sun and the 30 degree astrological signs

SAGITTARIUS   SCORPIO

CAPRICORN   LIBRA

AQUARIUS   VIRGO

PISCES   LEO

ARIES   CANCER

Orion

TAURUS   GEMINI

PRECESSION OF THE EQUINOXES

This is the definition of precession from Britannica.com:

Motion of the equinoxes along the ecliptic (the plane of the Earth's orbit) caused by the cyclic precession of the Earth's axis of rotation.

The most practical way of describing this phenomenon is to ask you to imagine a child's top as it starts to slow down, it starts to wobble in a direction opposite to its direction of spin. This is exactly what the Earth does as well as other planets, only in our case it takes thousands of years to complete one revolution of precession. Precession affects all spinning objects including the sun, which is estimated to have a precessional angle of five degrees. It is current belief that the precession of the equinoxes is caused by the action of the gravitational pull of the sun and moon upon the equator of the Earth and astronomers give this phenomena the name of *"Lunisolar Precession"*. However, whereas I agree that there must be some effect caused by gravity and the evidence of this is the bulge at the equator and the flattening at the poles.

I believe that science may be currently mistaken and that there is a further and more powerful engine at work. This engine may be associated with the recently discovered effects of Nuclear Magnetic Resonance and is a further explanation to some of the historical results that we see in myth, legend and science. The effect of NMR is that an atom precesses around its magnetic poles, and when a certain radio frequency is directed at that atom, it flips to either ninety or one hundred and eighty degrees. Different types of atoms recover to their original state of equilibrium at different rates, which allows images of the interior of solids or fluids to be seen by the observer conducting the experiment. This system is also used in modern medicine for body and brain scans.

Gravity is a weak force and can not be equated to the strength of magnetism. I believe that there is insufficient proof that a gravitational effect is the cause of precession to such a large body as the sun and therefore I consider that electro magnetic energy is the likely source of the phenomenon of precession. A knock over of 180° and no recovery could explain why the planet Venus is in an opposite rotation to the other planets. Perhaps the sun gives off occasional radio bursts of the right frequency to affect the core of a planet. Whereas I realise that microcosms such as atomic and particle universes are not necessarily compatible with macrocosms in terms of their physics, I consider that this is a theory well worth investigating.

The history of the discovery of precession is given by modern academics to the Greeks, despite the contrary evidence of Greek philosophers like Plato. In compiling his famous star catalogue (completed in 129 BC), the Greek astronomer Hipparchus noticed that the positions of the stars were shifted in a systematic way from earlier Babylonian (Chaldean) measures. This indicated that it was not the stars that were moving but rather the observing platform - the Earth. Such a motion is called precession and consists of a cyclic wobbling in the orientation of the Earth's axis of rotation with a period of almost 26,000 years. Precession was the third discovered motion of the Earth, after the far more obvious daily rotation and annual revolution. To a much lesser extent, the planets exert influence as well. One of his areas of study being the ancient measure, Sir Isaac Newton explained that the Earth's axis describes a circle with a diameter of 46.8° about the pole every 25,920 years.

The projection onto the sky of the Earth's axis of rotation results in two notable points at opposite directions: the north and south celestial poles. Because of precession, these points trace out circles on the sky. Today, the north celestial pole points to within just 1° of the arc of Polaris. It will point closest to Polaris in AD 2017. In 12,000 years the north celestial pole will point about 5° from Vega. Presently, the south celestial pole does not point in the vicinity of any bright star. Also moving with this wobble, is the projection onto the sky of the Earth's equator. This projection, a great circle, is called the celestial equator. The celestial equator intersects another useful great circle, the ecliptic. As the Earth orbits the Sun, the constantly changing direction from which we view the Sun causes it to trace out the ecliptic. The celestial equator is inclined at a 23.4° angle to the ecliptic (the so-called obliquity of the ecliptic). The celestial equator and the ecliptic intersect at two points called the equinoxes (vernal and autumnal). During the course of the year, as the Earth orbits the Sun, the latter is seen crossing the equator twice, in March moving from the Southern Hemisphere into the Northern Hemisphere and in September moving in the opposite direction. The equinoxes drift westward along the ecliptic. They drift at the rate of 50.2 arc-seconds annually as the celestial equator moves with the Earth's precession. Each cycle takes 25,920 years, and our ancestors divided the cycle of precession into twelve. The resulting twelve 2,160 year epochs were also named after the signs of the zodiac with which they measured the months of the year.

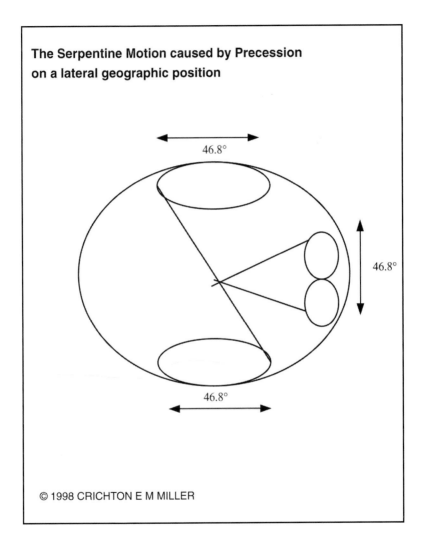

The Serpentine Motion caused by Precession on a lateral geographic position

46.8°

46.8°

46.8°

© 1998 CRICHTON E M MILLER

25,920 y
360 d = 72y

Today, the rate of the wobble (precession) is established as 1 degree per 72 years.

THE FIGURE OF EIGHT

I discovered this unusual figure of eight effect of the action of precession by experimenting with a specially designed globe. This globe is fixed on a rotating table at an angle of 23.4° and is also able to freely rotate on its own axis. The result of mimicking the actions of precession with the model was quite remarkable. It showed that any fixed position on the sphere forms or describes a figure of eight motion with a twist at the apex of the figure. From an earthly observation point, the effect would be to twist the perspective of stars and constellations being observed on the ecliptic plane. Robert Bauval and Adrian Gilbert, in their book *"The Orion Mystery"* have faced some criticism along with Graham Hancock in his book "Fingerprints of the Gods". The criticism levelled was on the basis that the three stars of Orion's belt did not align with the three pyramids of Giza and in point of fact were several degrees out. The phenomena that I have discovered shows that the twisting of perspective caused by the figure of eight motion may well explain this anomaly. However, I also feel that they were aligned in this way, was to emphasise the effects of the spring equinox. My research has showed that this pattern of offset is repeated in stone circles in other parts of the world. The sun lives in the house of each sign of the zodiac, at the spring equinox for 2,160 years, before moving into the next sign along the ecliptic. The vernal point is currently leaving Pisces and entering Aquarius. This places our epoch at the end of the precessional half cycle that started with the age of Leo. The rotation of the axis is shown by changing pole stars, Polaris and Vega, which are viewed at each end of the precessional half cycle of 12,960 years. Vega was the northern pole star in 10,500 BC during the age of Leo and Polaris is the current pole star of the Age of Pisces.

ECLIPTIC PLANE

In astronomy, the great circle that is the apparent path of the Sun among the constellations in the course of a year, or from another viewpoint, the projection on the celestial sphere of the orbit of the Earth around the Sun. The constellations of the zodiac are arranged along the ecliptic. The ecliptic is inclined about 23.4° to the plane of the celestial equator, the two points of intersection of the ecliptic and the plane mark the vernal and autumnal equinoxes.

In the ecliptic system of astronomical co-ordinates, celestial longitude is measured in degrees east from the vernal equinox along the ecliptic. Celestial latitude is measured in degrees north (positive) or south (negative) from the ecliptic to the ecliptic poles. Each ecliptic pole is 23.4° from the corresponding celestial pole.

Definition from Britannica.com:
> *The ecliptic is the angle of the obliquity of the Earth to the plane as it orbits around the sun on its annual voyage.*[xlvii] *The obliquity is caused by lunisolar precession.*

I believe that Hipparchus inherited the knowledge of precession from the Egyptian priests during the Ptolmic reign. My evidence for this comes from the concept of the Egyptian ages. In the annual cycle, the sun moves across the zodiac constellations at 1° every 24 hours and one constellation per month i.e. Aquarius followed by Pisces, Aries, Taurus, Gemini, Cancer and Leo. Precession however, is a different matter. The cycle causes the movement of the sun to slowly reverse on the monthly rate at 1° every 72 years. The result being to move the seasons forward one day every 72 years or one month every 2,160 years. Consequently, Orion would have been visible at midnight, on the meridian at the Autumn Equinox, at the time of building Khufu's Pyramid. Orion is now seen on the meridian at midnight of the Winter Solstice, giving further explanation to the ancient Egyptian saying: "*summer has become winter and winter has become summer and all is confusion*". This ancient proverb may have resulted from the decline of the priests and scribes or their knowledge of precession and the seasons.

The effect that would have been witnessed at the site of Great Pyramid of Khufu, had there been an observer there for the last half cycle of the precession is the one drawn as a conclusion by Robert Bauval in his book "*The Orion Mystery*".[xlviii] Robert appears to consider that the ancient observers would have maintained the calendar on the same date thus noticing that Orion was slowly rising in the sky due to the precessional motion. It is my belief that the Egyptians were far more advanced than that. That they were quite able to work out that the seasons were moving forward and that Orion was to be observed at the same altitude at different months of the year, over the ages, apart from when it was hidden by the sun. The cross and plumbline is able to show the effects of precession in a period of only nine years with careful observation. This will show a movement of 7.5 minutes of arc and well within the reaches of a basic instrument that is capable of a measurement of 3 arc minutes. It is certain using reason and common sense, that a difference of one degree would be noticed over a period of 72 years because one degree is a whole hour. These people kept the time for thousands of years with meticulous consistency.

The action of precession causes the sun, when rising at the spring equinox, to take 2,160 years to move across the 30° face of a zodiac sign. The direction of the movement of the sun is through the zodiac signs in the direction of Leo, Cancer, Gemini, Taurus, Aries, Pisces, and Aquarius. In our present age, we are about to leave Pisces and enter the Age of Aquarius; there is great controversy about when the actual date of the transit of the vernal point will be.[xlix]

The Roman Catholic Church and modern born again Christians use fish imagery to indicate the present astrological age of Pisces and one of Christ's messages was documented that he sent his disciples out to be Fishers of Men, while he had come as a Shepherd to save the lost Sheep. Prior to this occult knowledge practised by a church that forbids such things, was the evidence of the group known as the Essenes. The Essenes are the name given to the group at Qumran that wrote the Dead Sea Scrolls and this name translates from Arabic as "*School of Fishes*".

The age prior to Pisces was Aries, the sign of the Sheep or Ram. The iconography of the zodiac for this period is the same shape as the bronze top to the cross of the House of Amen. The Egyptian name for the sign of Aries was Amen. Evidence that they were aware of this age is seen in the naming of their Pharaohs. For example, take Tutankhamen, the boy pharaoh as an example. Translated, his name means "Life or reflection of Aries". It was fairly common practice in ancient times for a series of Pharaohs to be named after the age they were living in, e.g. Amenhotep I and Amenhotep II. Only the rebel Pharaoh Akhenaton (the son of Amenhotep III) who caused insurrection by changing the religion of the time toward Ra (the sun) as the giver of life broke this by changing his name.

Prior to the age of Aries, the Egyptians were seen to have revered Bulls and Cows, thus reflecting the age of Taurus. Evidence of this awareness can be seen in the worship of the Apis bulls. Herodotus wrote:

> '*The Apis bull is the calf of a cow, which is never able after to have another. The Egyptians believe that a flash of lightening strikes the cow from heaven and thus causes her to conceive the Apis. It has distinctive marks. It is black with a white diamond on its forehead, the image of an eagle on its back, two white hairs on its tail and a scarab beetle mark under its tongue.*'

The Egyptians believed that the Apis was the earthly incarnation of a god that actually lived amongst the human population and was thought to be the manifestation of Osiris. Upon death, the bull was treated with the same glory as a pharaoh. They were mourned and mummified and laid to rest with dignity. Plutarch describes the Apis as being '*a fair and beautiful image of the soul of Osiris*'. A further 2,160 years earlier, the age was Gemini (the sign of the twins). At this time Egypt was divided in two as the North and South Kingdoms under Set and Osiris. The age of Cancer was characterised by the ancient Egyptians as the Scarab. There is a well-documented argument that the Sphinx was symbolic of the age of Leo and was constructed to show the rising of the sun in the East at the Spring Equinox in that age. Certainly lions were revered from a remote period of time in Egypt. So we have more circumstantial evidence of the knowledge of precession stretching back at least six thousand years to the beginning of the age of Taurus.

NUTATION

The effect of nutation, or the cycle of the nodes, as described by Astronomers is "*a nodding of the Earth every 18.6 years of approximately 12 degrees*".[li] The gravitational effect of the moon causes this to occur. Did the ancients understand the effects of the phenomenon of nutation? The ancient peoples living in the British Isles and northern France certainly were aware of its occurrence as they built their stone circles for a reason more practical than religious. These astronomical stone circles are located in a large sweeping curve, up the westerly side of France and Britain and all along the Atlantic sea board. According to archeo-astronomers, they were mainly built to observe the eighteen year cycle of the moon nodes and were aligned to hilltops so that the moon could be seen to "*touch*" the top of the hill. The circles were places that were designed so as to be able to calculate the angles of the nodes and the standstill of the moon as well as other heavenly motions. Is it possible that this wobble of the planet is sufficient to cause weather changes that may occur on a cyclical basis in marginal areas? Changes, which might affect the food, supply? The modern concept of severe El Nino effects might be caused by nutation in combination with increased solar activity. It may also be the case that the story of Joseph in the Bible had a relationship to the understanding of astronomy and how astronomical changes can affect the environment.

Joseph was considered to have worn a multi coloured coat similar to the legends told of the rainbow-coloured serpent of the Maya and Inca peoples. Joseph's interpretation of the meaning of the seven fat cows and seven thin cows of the Pharaoh's dream that represented seven years of feast and seven of famine may have just fitted in well with the cycle, especially if Egypt was becoming more marginal as an agricultural area. There are many indications showing that the area around Egypt was originally good grazing land for sheep and goats and that this dried out and became desert due to the changes in rainfall and climate. Was Joseph able to warn the Pharaoh of the coming disaster through his knowledge of astronomy and cyclical effects, thus saving Egypt and his own people who relied on the grazing for their sheep? Certainly, although it is not a historical story, it is an indication that there was more to the story than originally meets the eye. As the ancients observed the sun, moon and planets at all the times and told the time by their motions, they would easily be able to notice the effect of nutation with the cross and plumbline that I propose they used.

THE CIRCUMSTANTIAL EVIDENCE

The purpose of this section of the book is to demonstrate that the simple technology required for producing and using such an instrument was in place at that time.

I am either a genius and have invented something never before seen by Man, or I am a simple researcher gifted with the skills to rediscover this lost instrument through unbiased observation. I believe it to be a case of the latter being true. Ancient peoples overcame problems, survived and flourished through innovation. They perfected their methods over thousands of years and left a legacy for us, that we still exploit to this day. Their workings demonstrate that they had to use some sort of device to measure and observe the movement of the sun, moon, planets and stars accurately. That something had to be of a simple nature to suit the already discovered artefacts of the times. To say it was anything other than 'simple' would lead to speculation taking hold in the public mind that they had a superior technology to us. They did not have a superior technology, but they did have a different type of knowledge. This knowledge was so alien to our modern minds that its physical manifestations leads some people to the firm conclusion that little green men from outer space assisted them. Considerable changes in the thinking of some of the public was instigated by such books as "Was God an Astronaut" written by Erich Von Deineken in the early 1970's. I also read this book and was nearly driven by such enthusiasm to find evidence of the aliens that I almost gave up my work and booked a one way ticket to Peru. No, the conclusion that I have come to over the years, is that *we* were the aliens and that we are the poor surviving children of a past civilisation wiped out by the cataclysm of the melting ice caps at the end of the last ice age. That we have rebuilt this civilisation on the fragments of knowledge left to us by our seagoing ancestors. The thing that we have not got together, that might end up in our own destruction again, is the understanding of our relationship with the oneness of nature and our understanding of time and place. We currently are a species mostly driven by material gain and ego and that will be our self destructive and divisive downfall.

I have now built the cross and plumbline in many forms incorporating all the basic principles outlined, and through much experimentation, as shall be illustrated later, I have proved that my instrument does work in reality. It is no longer a theory, but is now a fact. It is a fact that it will carry out the tasks claimed and many more besides because the application is only limited by the knowledge and ability of the operator. Any race of people who could build the great pyramids of the world must, by reason and common sense, have had great knowledge and an instrument to design them with. It is said that the Egyptian priests told Plato that the Greeks were as children compared to themselves and were amused by their naive questions. Think how far we have come in a hundred years and try to imagine a people who developed their knowledge over thousands of years.

The Plumb Line

The plumb line and bob are the keys to the instrument. The line is essential and the bob is the weight attached to the line which always points to the centre of the Earth due to the science of gravitation. The law of gravitation creates a constant for the measuring device. The plumb line is seen as an agricultural innovation developed from the scales used for measuring grain and other agricultural products and manifests itself as a levelling device on many forms of recorded artwork from the ancient Egyptians. The "Weighing of the Heart" ceremony in the funerary rites, being a prime example.*[lii]*

It would seem that we do have an echo of the original 'cross' as used by the ancient Egyptians and the Overseer of the Hour.*[liii]* It was an instrument called the Merkhet and it seems to have been in use from about 1500BC onwards. Essentially this instrument was a stick with a notch in the end for sighting and a plumb line (nothing more than string with a weight). The Merkhet and a plumb line can be used to measure a star position in terms of compass direction and slope of a sight line. However, this measurement only works for the location you are in, an astronomer with a Merkhet in a different location would get a different measurement. Could it be that the Merkhet was an 'echo' of the original cross? Certainly the general idea is along the same lines, but the execution is less sophisticated and the technology employed less advanced. The Merkhet was probably influential in the design of the later Roman Groma, the principle surveying instrument of the Roman Agrimensores, the land surveyors. Simple in design, the instrument consisted of a set of crossed arms resting on a bracket and attached to a vertical shaft. Each of the four arms had a cord with a hanging plumb bob. Designed to survey straight lines and right angles, it was nevertheless pretty useless in a high wind. There is a replica of a Groma in the London science museum. Do the Merkhet and Groma prove a descent, a lineage of design and technology? I certainly believe so and I believe that the precursor to these instruments was the Cross. The only missing ingredients in modern images of the cross that are seen on churches and gravestones, are the plumb bob and plumb line.

The Staff

Staffs, surveying rods used for the measuring of heights and distances, are commonly seen in Egyptian Art. The notched palm rib, which was used as an early theodolite, was known as a Bay.[liv]

The Cross Arm

Cross arms are seen in many applications. Notably observed in weights and measures.

Dividers

It is widely accepted that the ancient peoples had perfected the mathematical science of geometry; the instrument known as dividers is an essential tool for the application of this skill. There is the image of a relief of Thothmosis in the Heb Sed Festival, depicted with something that appears to be an instrument in his hand - is it possible that this instrument could in fact be a set of dividers?

The Measure

There were probably two types of measurements employed by the ancient Egyptians. One measurement was designed for scientific purposes and the other for agricultural uses. David Ritchie describes a form of measure later on in this book with his work on the Stone of Destiny. The agricultural system of measurement was employed mostly for the pacing out of field areas and general measurements requiring only a reasonable accuracy. This was essential after the inundation of the Nile where farmers' boundaries were swept away by the flood and had to be re-established. The types of measurements used were most likely to be similar to the British imperial measures, which were inherited from the Romans, such as yards, feet and inches. This form of measurement has sadly been lost in Europe and the United Kingdom as the governments changed to the metric system. The younger generations of Europe do not even remember it as it has been bred out of their knowledge. It is still being used in America and that is its last place of preservation to my knowledge. For a farmer working out land area for planting or ownership, then feet and yards would be useful, such as one-pace equals three feet and three feet equal one yard. However, the Egyptian architects and astronomers used a higher level of measurement for the construction of their temples, known to be in cubits and Royal cubits and divided into decimal points in the fourth Dynasty.

Degrees

The degree is a form of measurement that was employed by the ancient Egyptians and made possible with decimalization, as shown by the research of Sir William Flanders Petrie. Degrees have since become associated with a designation of rank in the Masonic hierarchy e.g. thirty-third degree mason, and in the designation of accredited learning issued by academic institutions.

Paper

The Egyptians invented a paper known as papyrus and used it in the form of scrolls for many forms of writing. Papyrus was a material, which they were able to record time and astronomical measurements on and many friezes show Thoth recording and writing on this material.

The mathematical knowledge required

The main mathematical system, in my opinion, was practical geometry. The formula of Pi is considered by many, to be incorporated in to the construction of the Pyramid of Khufu. Referral to Sir William Petrie's work, when he surveyed the pyramids, confirms this intensive and skilled knowledge.

Astronomy

It is well known that the ancients practised astronomy as a way of life and the references provided in this book uphold that view without much argument. My research is beginning to reveal that they had a greater understanding than previously imagined.

Timekeeping

Horology is the correct name for time keeping and the name is devised from the Egyptian god Horus, who was well known as the timekeeper.

Conics

Conics is the use of focal points and reflection, which can be seen clearly designed in the pyramid constructions with their polished surfaces, in water observatories and is a form of geometry. The Amerindians speak of a smoking mirror with which an individual can observe the workings of man and the gods. Remember that the ancients saw the stars and sun as gods.

Written evidence of the use of an Inclinometer in Prehistoric Times

The following is an extract that clearly shows the various skills of astronomical measurement that I claim was being used in ancient times. This research by Simon Cox displays an ancient Coptic story on the measurement of the angles of stars. The following account also comments on the disaster of the Flood, the art of prophecy through astronomy/astrology and the gravitational effects on a population caused by a sudden tilt of the Earth. The effects of nuclear magnetic resonance instigating a continental crustal slip on the Earth long ago, may have caused the sudden tilt that is described.

In Volume two of 'Pyramids of Giza' by Col Howard Vyse, 1842, the following commentary on some Coptic legends is related:

> *"Masoudi's account professes to relate the Coptic tradition, which says, 'That Surid, Ben Shaluk, Ben Sermuni, Ben Termidun, Ben Tedresan, Ben Sal, one of the kings of Egypt before the flood, built the two great pyramids; and, notwithstanding they were subsequently named after a person called Sheddad Ben Ad, that they were not built by Adites, who could not conquer Egypt, on account of the powers, which the Egyptians possessed by means of enchantment; that the reason for building the pyramids was the following dream, which happened to Surid three hundred years previous to the flood. It appeared to him, that the Earth was overthrown, and that the inhabitants were laid prostrate upon it; that the stars wandered confusedly from their courses and clashed together with a tremendous noise. The king, although greatly affected by this vision, did not disclose it to any person, but was conscious that some great event was about to take place. Soon afterwards in another vision, he saw the fixed stars descend upon the Earth in the form of white birds, and seizing the people, enclose them in a cleft between two great mountains, which shut upon them. The stars were dark, and veiled with smoke. The king awoke in great consternation, and repaired to the temple of the sun, where with great lamentations, he prostrated himself in the dust. Early in the morning he assembled the chief priests from all the Nome's of Egypt, a hundred and thirty in number; no other persons were admitted to this assembly, when he related his first and second vision. The interpretation was declared to announce, 'that some great event would take place'.*

The high priest, whose name was Philimon or Iklimon, spoke as follows:- 'Grand and mysterious are thy dreams: The visions of the king will not prove deceptive, for sacred is his majesty. I will now declare unto the king a dream, which I also had a year ago, but which I have not imparted to any human being'. The king said: 'Relate it O Philimon' the high priest accordingly began: 'I was sitting with the king upon the tower of Amasis. The firmament descended from above till it overshadowed us like a vault. The king raised his hands in supplication to the heavenly bodies, whose brightness was obscured in a mysterious and threatening manner. The people ran to the palace to implore the kings' protection; which in great alarm again raised his hands towards the heavens, and ordered me to do the same; and behold, a bright opening appeared over the king and the sun shone forth above. These circumstances allayed our apprehensions, and indicated that the sky would resume its former altitude; and fear together with the dream vanished away.'

The king then directed the astrologers to ascertain by taking the altitude whether the stars foretold any great catastrophe, and the result announced and approaching deluge. The king ordered them to inquire whether or not this calamity would befall Egypt; and they answered yes, the flood will overwhelm the land, and destroy a large portion of it for some years.

He ordered them to inquire if the Earth would again become fruitful, or if it would continue to be covered in water. They answered that its former fertility would return. The king demanded what would then happen. He was informed that a stranger would invade the country, kill the inhabitants, and seize upon their property and that afterwards a deformed people, coming from beyond the Nile, would take possession of the kingdom; upon which the king ordered the pyramids to be built. The predictions of the priests to be inscribed columns, and upon the large stones belonging to them; and he placed within them his treasures, and all his valuable property, together with the bodies of his ancestors. He also ordered the priests to deposit within them, written accounts of their wisdom and acquirements in the different arts and sciences. Subterraneous channels were also constructed to convey them to the waters of the Nile. He filled the passages with talismans, with wonderful things, and idols; and with the writings of the priests, containing all manner of wisdom, the names and properties of medical plants, and the sciences of arithmetic and of geometry; that they might remain as records, for the benefit of those, who could afterwards comprehend them.

He ordered pillars to be cut, and an extensive pavement to be formed. The lead employed in the work was procured from the west. The stone came from the neighbourhood of Es Souan. In this way were built the three pyramids at Dashoor, the eastern, western, and the coloured one. In carrying on the work, leaves of papyrus, or paper, inscribed with certain characters, were placed under the stones prepared in the quarries; and upon being struck, the blocks were moved at each time the distance of a bow shot (about one hundred and fifty cubits), and so by degrees arrived at the pyramids. Rods of iron were inserted into the centres of the stones, that formed the pavement, and, passing through the blocks placed upon them, were fixed by melted lead. Entrances, with porticoes composed of stones fastened together with lead, were made forty cubits under the Earth: the length of every portico being one hundred and fifty cubits. The door of the eastern pyramid was one hundred cubits eastward from the centre of the face, in which it was placed, and was in the building itself. The door of the western pyramid was one hundred cubits westward, and was also in the building. The door of the coloured pyramid was one hundred cubits southward of the centre, and was likewise in the building. The height of each pyramid was one hundred royal cubits, equal to five hundred common cubits and the squares of the bases were the same, they began at the eastern side. When the buildings were finished, the people assembled with rejoicing around the king, who covered the pyramids with coloured brocade, from the top to the bottom, and gave a great feast, at which all the inhabitants of the country were present.

He constructed, likewise, with coloured granite, in the western pyramid, thirty repositories for sacred symbols, and talismans formed of sapphires, for instruments of war composed of iron, which could not become rusty, and of glass which could be bent without being broken. Also for many sorts of medicines, simple in compound, and for deadly poisons. In the eastern pyramid were inscribed the heavenly spheres, and figures representing the stars and planets in the forms, in which they were worshipped.

The king also deposited the instruments, and the thuribula, with which his forefathers had sacrificed to the stars. Also their writings; of the positions of the stars, and their circles; together with the history and chronicles of time past, of that, which is to come, and of *every future event, which would take place in Egypt. He placed there, also, coloured basins (for lustration and sacrificial purposes), with pure water, and other matters.*

Within the coloured pyramid were laid the bodies of the deceased priests, in sarcophagi of black granite; and with each was a book, in which the mysteries of his profession and the acts of his life were related. There were different degrees among the priests, who were employed in metaphysical speculations, and who served the seven planets. Every planet had two sets of worshippers; each subdivided into seven classes. The first comprehended the priests, who worshipped, or served seven planets. The second, those who served six planets; the third, those who served five planets; the fourth, those who served four planets; the fifth, those who served three planets; the sixth, those who served two planets; the seventh, those who served one planet. The names of these classes were on the sides of the sarcophagi; and within them were lodged books with golden leaves, upon which each priest had written a history of the past and a prophecy of the future. Upon the sarcophagi were, also, represented the manner, in which arts and sciences were performed, with a description of each process, and the object of it. The king assigned to every pyramid a guardian.

The guardian of the eastern pyramid was an idol of speckled granite, standing upright, with a weapon like a spear in his hand. A serpent was wreathed around his head, which seized upon and strangled whoever approached, by twisting round his neck, when it again returned to its former position upon the idol.

The guardian of the western pyramid was an image made of black and white onyx, with fierce and sparkling eyes, seated on a throne, and armed with a spear; upon the approach of a stranger, a sudden noise was heard, and the image destroyed him.

To the coloured (that is, third pyramid) he assigned a statue, placed upon a pedestal, which was endowed with the power of entrancing every beholder till he perished.

When everything was finished, he caused the pyramids to be haunted with living spirits; and offered up sacrifices to prevent the intrusion of strangers, and of all persons, excepting those, who by their conduct were worthy of admission. The author then says, that, according to the Coptic account, the following passage was inscribed, in Arabic, upon the pyramids. 'I, Surid, the king, have built these pyramids, and have finished them in sixty-one years. Let him, who comes after me, and imagines himself a king like me, attempt to destroy them in six hundred. To destroy is easier than to build. I have clothed them with silk; let him try to cover them with mats'."

Masoudi, who supposedly related the above story from a Coptic original, died around 957 AD. He was sometime known as the 'Herodotus of Iran' on account of his combing history and large scale geography in a single work. It seems he travelled extensively in Persia, India and throughout the Middle East, eventually settling in Egypt where he died after having written a thirty volume book of his journeys.

Another Arab historian, Kodhai, also relates a curious paragraph from another Coptic manuscript, supposedly found in the Monastery of Abu Hormeis. After having related the story of the finding of the manuscript, Kodhai then translates:

"In this manner were *the pyramids built. Upon the walls were written the mysteries of science, astronomy, geometry, physics, and much useful knowledge, which any person, who understands our writing can read. The deluge was to take place when the heart of the Lion entered into the first minute of the head of Cancer, at the declining of the star. The other indications were the sun and the moon entering into the first minute of the head of Aries and Saturn, in the first degree and twenty-eight minutes of Aries. Jupiter, in the twenty-ninth degree twenty-eight minutes of Pisces; and Hermes, (Mercury), in the twenty-seventh minute of Pisces; the rising Moon, in the fifth degree and three minutes of the Lion"*

Here we have an account researched by Simon Cox, that describes measurements of the altitude of the stars. The story also explains the likely gravitational effects of crustal slip or pole shift as one might expect if nuclear magnetic resonance theory held good and caused the destruction of the Earth in the time of the Flood. There are many other points worth noting such as measurements of the planets and stars to an accuracy of minutes. How did they make these observations to an accuracy of minutes? A telescope? It is possible, as lenses milled from crystal have been found recently, belonging to the Vikings and date as being over a thousand years old. It is also considered that the crystals found originated in Eastern Europe.[iv] But the telescope is linear operation and does not take sidereal (constellation) angles of measurement. Sidereal or horizontal measuring capacity with an exponential scale rule, is an essential and unique feature of the instrument and a necessity for the measurement of the stars. If the answer was not through the use of a telescope, how was it proposed that the ancients made these calculations? The answer, in my mind, is a simple product of two crossed staffs, a plumb line, a scale and a lot of knowledge.

All these skills can be demonstrated to have existed in some form or other from the Neolithic to the Pyramid age. Therefore, there can not be a reasonable argument put that they were not able to use the cross and plumbline because of the lack of the skill and materials required. So, on the supportive side of one half of the argument that the cross, plumbline and the knowledge could exist, is the clear evidence that the materials were available to the ancients. On the other half of the supportive argument, is the physical evidence that the ancients actually built the pyramids, stone circles and henges, practised astronomy, astrology and produced geometrical mathematics. More evidence of this ability to measure the angles of the stars and planetary bodies is revealed in the infamous Dead Sea Scrolls that were left in Qumran by the departing Essenes.

THE DEAD SEA SCROLLS

Found in 1947, the so-called 'Dead Sea Scrolls' have engaged debate and criticism ever since. The exact timetable of their finding is still not known, but from what researchers have pieced together we can deduce that a Bedouin shepherd found at least seven scrolls. They were concealed and protected in jars, in caves high above the settlement of Khirbet Qumran on the shores of the Dead Sea possibly in February of 1947. This discovery was against the turmoil and political backdrop that was Palestine in the late forties. The scrolls are then taken to a Bethlehem antiques dealer called Kando in the April of 1947, where he manages to purchase four of them, including the Isaiah Scroll, the Habakkuk Commentary, the Genesis Apocryphon and the Community Rule. Kando then sells these on to the Syrian Orthodox Archbishop of Jerusalem, Athanasius Yeshue Samuel. Another Bethlehem antiquities dealer called Feidi Salahi somehow manages to get hold of two more of the scrolls by November of 1947 and shows them to Hebrew University Professor Eliezer Sukenik. By the next month in December of 1947, Sukenik has himself managed to purchase three scrolls, another Isaiah Scroll, the War Scroll and the Hodayath. All it seems bought from Feidi Salahi. Sukenik arranges to view the scrolls previously purchased by Archbishop Samuel, but fails in his attempt to buy them from him. Meanwhile a copy of the Isaiah Scroll is shown to scholar John C. Trevor of the American Schools of Oriental Research (ASOR) Centre in Jerusalem, who along with colleague William Brownlee, photographs and identifies it. Suddenly the race was on to find more of the scrolls and to place them within a historical context. Trevor calls in renowned archaeologist William F Albright to confirm the identity of the scroll he had photographed, Albright does so and tells Trevor that in his estimation this is the oldest known Hebrew manuscript.

The story of the recovery of all the scrolls and of their background is a long one and can be found in many books on the subject, but here I will briefly give a description of some of the scrolls that were found.

The Scrolls from:

Qumran Cave One:

The Great Isaiah Scroll
Undoubtedly the oldest complete manuscript of Hebrew scripture yet found, with a date of pre 100BC.

The Second Isaiah Scroll.
This is a damaged scroll containing nearly a third of the text of Isaiah, in its wording it generally follows the Hebrew bible.

The Community Rule
A large six-foot scroll of some eleven columns found almost intact. This contained instructions for initiates into a hierarchical covenant community, or Yahad, of the 'Sons of Light', led by Zadokite priests who were separate from the 'Sons of Darkness' to purify themselves in preparation for the impending last judgement. Fragments of ten other copies of this scroll were found in caves four and five.

Pesher, or interpretation, of Habakkuk
Measuring some four and a half feet with thirteen columns, this scroll interpreted the Hebrew prophet as envisioning the defeat of a 'Wicked Priest' who 'ruled over Israel', but defiled himself by amassing wealth whilst robbing 'the poor' and persecuting an exiled 'Teacher of Righteousness'.

The War Scroll
A nine feet six inch scroll composed of some nineteen columns. This scroll was conceived as a sequel to the eschatological war envisioned in Daniel 11-12. Detailed instructions for the battle are given, in which the 'Sons of Light' are to massacre the 'Kittim', or the 'Sons of Darkness' led by Satan.

Qumran Cave Three:

<u>The Copper Scroll</u>
An amazing list of buried treasure. The Copper Scroll details certain sites throughout Judea where gold and silver treasures are buried. Originally connected as three copper sheets, the scroll was found as two separate rolls at the back of the cave. The text has taken some time to be deciphered and this scroll remains at the centre of heated scholarly debate over its meaning and origins. So far, archaeology has failed to locate any of the treasure.

Qumran Cave Four:

<u>The Book of Enoch</u>
Previously only known in translation, these fragments of the Book of Enoch contain elements of the Book of Watchers [1 En 6-36], the Astronomical Book [1 En 72-82], Dream Visions [1 En 80-90] and the Letter of Enoch [1 En 37-72].

Qumran Cave Eleven:

<u>The Temple Scroll</u>
Text containing idealised description of a temple with huge courts (as large as Jerusalem), from which all of the impure, including gentiles, are excluded. Also describes a sacrificial cult using a solar calendar.

Various methods used in dating the ancient documents have found that most of the scrolls were written, copied or composed between the third century BC and the first century AD. The Priestly Courses onward and the calendrical text show accurate reading of the lunisolar calendar.[lvi] In Brontologion of the Dead Sea Scrolls written by the Qumran community by the Essenes, the art of Astrology and Astronomy is clearly shown.[lvii] In Fragments 1 column (5) to Fragment 2 Column (20) accurate positioning of the moon are observed against the Zodiac Signs. It is clear that they operated a 364-day year and that accurate astronomy was involved. But how did they measure the altitudes of the stars and the moon? The only way was with the cross and plumbline and I will now show you how latitude may be measured with this instrument.

CHAPTER 7

GODS BROUGHT TO EARTH

How To Find Latitude with the Cross

Calculating latitude using this system is an easy task, providing the sky is clear and one knows how to find the current pole star at night. Today, the North Pole star is Polaris and it can be found by following the pointer stars on the outer edge of the Dipper or Plough.

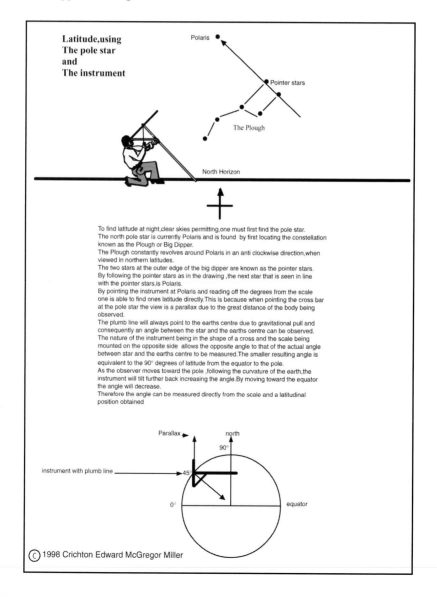

Latitude,using
The pole star
and
The instrument

Polaris

Pointer stars

The Plough

North Horizon

To find latitude at night,clear skies permitting,one must first find the pole star.
The north pole star is currently Polaris and is found by first locating the constellation known as the Plough or Big Dipper.
The Plough constantly revolves around Polaris in an anti clockwise direction,when viewed in northern latitudes.
The two stars at the outer edge of the big dipper are known as the pointer stars.
By following the pointer stars as in the drawing ,the next star that is seen in line with the pointer stars,is Polaris.
By pointing the instrument at Polaris and reading off the degrees from the scale one is able to find ones latitude directly.This is because when pointing the cross bar at the pole star the view is a parallax due to the great distance of the body being observed.
The plumb line will always point to the earths centre due to gravitational pull and consequently an angle between the star and the earths centre can be observed.
The nature of the instrument being in the shape of a cross and the scale being mounted on the opposite side allows the opposite angle to that of the actual angle between star and the earths centre to be measured.The smaller resulting angle is equivalent to the 90° degrees of latitude from the equator to the pole.
As the observer moves toward the pole ,following the curvature of the earth,the instrument will tilt further back increasing the angle.By moving toward the equator the angle will decrease.
Therefore the angle can be measured directly from the scale and a latitudinal position obtained

Parallax

north

90°

instrument with plumb line

45°

0°

equator

Practical Applications and Experimental Results

At the time of the building of the Great Pyramid of Khufu, the pole was in the constellation Draconis the Serpent and no direct star was above the pole. The shifting of the polestars is caused by the effects of precession and in my opinion, the Egyptians were well aware of this phenomenon. Their astronomers were certainly able to identify the area of the North Pole by using the circumpolar stars that drifted around the celestial pole as a grid reference.

An observer sighting along the cross arm at the area of the pole[lviii] would have to tilt the whole instrument backward towards them using the upright as a support. The tilt of the instrument increases the further north the observer travels, until if the North Pole were reached, the angle read from the scale would be 90° latitude and the star would appear directly overhead. If the observation was made at the equator, then the angle would be 0° latitude and the star would appear on the northern horizon. The different latitudes in between 90° and 0° are read from the point that the plumb line crosses the scale marking the degrees. The size of the triangle formed between the cross arm and the scale will determine the accuracy. Consequently, any form of measurement will suffice, as long as it is decimal based and equals 90 units and fractions thereof. Metres, inches or cubits will work to measure degrees and minutes of arc, providing the scale is large enough. This is the basic principle of a quadrant, from which an astrolabe is a descendant. The difference between a quadrant and the full working version of the Cross, is the extension of the upright or staff, allowing sidereal observations to be used and a fulcrum to be added. It had previously been thought that the ancients were capable of limited observational ability, when determining astronomical angles for the alignment of their pyramids and for astronomy. However, with this type of instrument, the materials for which were readily available, all observations were possible dependent upon the size of the instrument employed. By sighting the north star it follows that the reverse observation will give true south. True east and west are a logical progression. It then follows that the equinoctial points of the ecliptic on the horizon can be observed and the exact place of the rising of the sun and constellations during this biannual event can be marked for future observation.(See appendix 1)

Taking Lunar and Stellar Sights

By sighting the stars in conjunction with the appropriate almanacs and timepiece, it is possible to find both latitude and longitude. By first finding true north by sighting the pole star (in the northern hemisphere), true south may be obtained, as the reverse end of the cross arm will be pointing exactly true south during this observation. As all the heavenly bodies on the ecliptic travel from east to west in an orderly and predictable fashion, achieving given ascension and declination angles at particular locations and times. It is, therefore, possible to obtain longitudinal positions from the correct observations coupled with an almanac, and knowledge of local time in relation to world time at the prime meridian. The timepiece is obvious but no one seems to have cottoned on yet to the simplicity of this system. Due to the Earth's orbit around the sun, these stars move westward every night at the same observation time by 0.98°. In other words, their position over a given point on the Earth's surface moves westward 58.8 nautical miles. At the same time, as the Earth progresses around the sun and the seasons change due to the Earth's angle of obliquity to the ecliptic, the azimuth of the stars also changes in ascension or declination. Midwinter (solstice) is the maximum ascension of stars on the meridian north of the equator, in the night sky and midsummer is the minimum declination of the ecliptic stars.

Consequently, if a given star is predicted by the almanac to be at an angle of 23.4° above the equator and this same star is observed exactly due south at 0° longitude at 24.00 hours GMT, it becomes a relatively simple exercise to find your latitude position on the Earth's surface, using the instrument to within 3 to 6 nautical miles. The same is true of the sun because it is always on the ecliptic. The moon is a different matter because it crosses the ecliptic during its orbit.

(a) The observer must find latitude using the sight taken of the pole star.

(b) The observer finds true south.

(c) The observer looks for the projected ascension at midnight of the star, and deducts the latitude made good.

(d) The observer then turns the instrument 90° so that the instrument is at right angles to north and south with the cross arms pointing east and west.

(e) The observer can now measure the difference in degrees between the observed position of the star against its predicted position at 0° longitude.

(f) The observer then corrects the difference between the predicted angle of the star and the actual angle into degrees and minutes east or west of 0°.

How the Ancients Measured the Earth

The universality of this instrument comes into its own with sidereal observations. To calculate the angle of an object such as a hill, the angle of a pyramid or the angle of stars, planets and constellations, the instrument is turned sideways and observed from the front. An assistant may have had to hold it while the observer directed the angle to be calculated and read the scale. It is during astronomical observations that the bronze or copper fork mounted at the top of the upright on the Egyptian cross of Amen could be fully utilised. By viewing a circumpolar star or a star that was close to the southern horizon through this slot or sight, the angle from the star to the horizon or the meridian could be calculated in degrees and arc minutes by reading the plumb line against the scale.

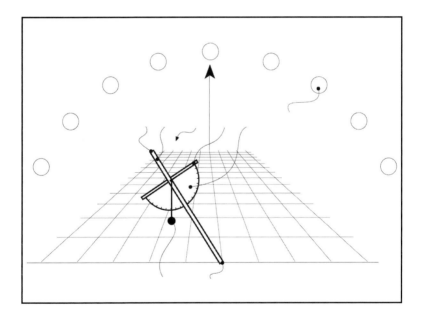

It is unlikely that sidereal observations were used on the ecliptic stars at the latitude of Giza, however those on the southern and northern horizons could be measured effectively with this method. The "imperishable stars" of the north polar sky could be measured in this way, as could the belt of Orion and Sirius. The bronze fork attached to the top of the cross that was found in the Great Pyramid was made in a reflective material. So if a fire or torch was behind the observer, the dim reflection of the fire would highlight the fork which acted as a sight without obscuring the star being observed.

As I have already stated, 4500 years ago, there was no true pole star around which the axis of the earth rotated. However, if a star was to be viewed from within the north facing 26° shaft of the pyramid of Khufu, then that star would be a rotating polar star, so we know that shaft would be pointing 4° south of true north. But to achieve that shaft construction in the first place would require a line or rule to be employed between the rotating polar stars. The cross bars on the cross and plumbline are able to achieve this very effectively. The Celtic cross is even more efficient in this area of observation as you will see. By holding the circle of the cross up to the polar stars as they rotate about the invisible pole (of 4500 years ago) and the inside of the wheel of the cross, it is possible to identify the polar center as the center of the cross itself. Many astronomers and Archaeologists (such as Kate Spence of Cambridge University), argue about which stars were the ones that the line was stretched across to find true north. This is unimportant unless dating is an issue, because the principle is the same regardless of the stars involved. It takes patience, observation and the cross and plumbline. True north can be found by observing the sun of course by measuring its elevation before and after noon and this requires a cross and plumbline as well.

The logic follows therefore, that to measure degrees in the form of latitude, the observer must understand that the planet is a sphere. The type of measurement used, in this case degrees, on a north south axis, will give a starting point for determining the size of the sphere. An observer only has to travel north or south along the meridian of measurement, taking regular observations using any form of measurement, such as inches, feet, yards, cubits or meters. A distance can be determined for 1 degree quite easily without travelling the full distance of a whole degree, if the cross and plumbline is large enough. 10 minutes of arc, or 10 sea miles would be sufficient on a six foot model and the north south axis could be measured as 60 x 360 = 21600 nautical miles circumference.

We have no evidence that the ancients were aware of the earth's shape as being an oblate spheroid, so we must follow the concept that they did not, for this exercise. It is reasonable to assume that having considered this measurement to be accurate, that the observers would then translate this knowledge to the earth's equator to measure an east west distance at the latitude of zero.

There is no evidence that they were aware of equatorial bulge, so the measurements stand based on the shape of a sphere and they are reasonably accurate to this day. However, the east west circumference measurements of a sphere reduce as the lines of latitude move away from the equator toward the poles causing the measurements for the rotational distance and speed to become reduced as well.

This leads us into measuring the earth's circumference by first discovering its speed of rotation.

At the equator, 360° = 21600 sea miles and it takes 24 hours to complete one full rotation, which can be observed by the position of a star or the sun as it reaches true south every 24 hours: 360 ÷ 24 = 15. The result is that the earth rotates through 15 degrees or 900 sea miles every hour at the equator.

To see the way that they thought in geometric terms, draw a circle and divide it into two parts with an equator, forming a pyramid with 45° slopes with its apex at the pole and the base on the equator. The result is an imaginary cone, with the side of the pyramid forming an exponential scale similar in design to the one in the great pyramid.

Chart for rotational speed at degrees of latitude for the pyramid shape

Degrees of Latitude	Speed of Cone rotation / nautical miles per hour
90	0
80	100
70	200
60	300
50	400
45	**450**
40	500
30	600
20	700
10	800
0	**900 (equator)**
10	800

It is simple, the half way mark between 0° and 90° is 45° and at that latitude all you have to do is add a factor of ten to the degrees. QED, the earth spins at 450 nautical miles per hour at 45° on an exponential scale based on the (cone) pyramid shape.

The outer curve of the sphere now forms a bow shape where the distance between the exponential scale formed by the imaginary pyramid and the outer edge of the sphere is at its furthest at the latitude of 45°.

This distance reduces either side of 45° until it is equal at the pole and the equator. To find the correct speed at that latitude, we have to use cosines.

45 Cosine 0.7071067 x 900 = 636.39 sea miles per hour

So, at what speed does the earth rotate at the 30° latitude of Giza?

30 Cosine 0.866025 x 900 = 779.42 sea miles per hour.

Measuring the Earth's Circumference on a Line of Latitude using the Number of the Beast 666.

We are able to say that the ancient mariners may have used 60 nautical miles to equal 1° as we still do today. There are 360° around the earth taking 24 hours to make one full rotation. That it takes one hour (Horus) or 60 minutes (Minute parts of Horus) for the earth to rotate through 15 degrees at the equator, means that the earth rotates at 900 nautical miles per hour at that latitude. The Equator is established as being on latitude of 0°. 360 x 60 nautical miles when the latitude is measured at the equator, equals roughly 21,600 NM and is the approximate theoretical distance of the earth's circumference. NASA measures it today in meters and takes the equatorial bulge into account to give a finding of 40024800 meters

First I want to demonstrate to you what Revelations in the Bible may have been getting at with the number of the beast as 666. It will lead the seeker into mathematics, and sacred geometry. To find the increments for measuring the earth at different lines of latitude we have to interpolate the degrees between the equator and the pole.

If the imaginary pyramid we discovered is reflected south of the equator, it forms a square within the circle that is similar to the concept spoken about by the Egyptians as squaring the circle. So when we divide the 60 NM which is equal to the distance of a degree at the equator, with the 90° from the equator to the pole, we come up with a surprise. 60 ÷ 90 = 0.666, therefore 0.666 is equal to one degree in this form of measurement. It follows therefore that 10° are also equal to a divisive and incremental figure that is equal to 6.66. When you deduct the angle of the sun on ecliptic at its highest altitude of 23.4° at the summer solstice from the latitude of Giza, which is 30 degrees you are left with 6.66°. This is equal to 399.6 sea miles and shows that the Giza pyramid may be telling us its distance from the ecliptic maximum and the Valley of the Kings.

It makes sense that the Pharaohs would want to be interred after their deaths at the same latitude that the stars of the zodiac reached every winter solstice since they considered themselves to be a star in earthly form. If you also continue to consider the significance of this ancient knowledge and subtract the 23.4° from 90° of the angle between the equator and the poles, the answer is 66.6°. The Pyramid is telling us again about the biblical "Beast" which the ancient Egyptians called "Geb", and I believe that the beast is the earth. As the Egyptians have made quite clear to all scholars, Geb was the earth and Nut was the sky goddess. The ancient Egyptians knew themselves to be "Children of Geb" or Earthlings and they did not only see the Earth as a mother and nurturer, but also as a beast. Why should they take this view?

Is it because they were aware of its awesome power and capability to destroy mankind with one little geological upheaval on even a local scale or was it just the utter ruthlessness of its system of eat or be eaten? We are so arrogant in our seemingly safe world and think we can actually affect it in the long term with our puny ways and endeavors. We modern humans are like ants in a fire grate and have no real understanding of our tenuous existence. Only those who have stood in the crater of a volcano and imagined it during eruption have any idea of our true place in the world. We seem incapable of understanding that we live on a ball of fire that thunders around the solar system with such kinetic energy and electro magnetic power that we are totally insignificant. We live on a thin layer like the skin on a pot of porridge, which rests insecurely on top of a boiling cauldron of spinning molten magma that can spill out at any time and shatter our illusions. On the land, the crust is only 24 miles thick and in the oceans it is as thin as 6 miles.

Sometimes, volcanic explosions like Thera, Pompeii, Etna or Mount St Helen's destroy us and awe us for a while, but after a few generations we soon slip back into our complacency and arrogance. Occasionally there are super volcanoes like the one under Yellow Stone Park in America, that should it explode, would destroy all life in the Northern Hemisphere. But when there is a polar shift like the one that ended the ice age and caused The Flood, all life on a planet wide basis may be destroyed. The Ancients remembered this event and called the earth a beast and knew that Mankind's material manifestation in it was a desperate struggle for survival, from the cradle to the grave. Some civilizations were so terrified of the capabilities of the "*spirits of nature*" that they performed human sacrifices to appease them so that such an event as the Flood would not happen again. We have records that the Minoans started sacrificing children shortly after the eruption of Thera. The House of Ram left this knowledge about the earth and its true ruthless nature in stone and word. They left it for us to rediscover and the man that is associated with the beast may have been Khufu, for it is his Horizon and therefore, his number. These ancestors measured the beast and they called it that because it traps us in its material form and it cyclically destroys our material form as individuals and as civilizations.

So how did they get the measure of the Beast using the number 666? Let us measure the distance of the circumference of the imaginary pyramid within the earth at any latitude using the number of the beast, which is 0.666. Look at the latitude of 1°, this is the first degree north of the equator. We know that one degree is equal to 60 nautical miles so we are now 60 miles north of the equatorial latitude. The reason we want to find the circumference of the earth at that latitude is so that we know the distance of a mile in degrees east or west. It is obviously going to be less than the distance of a degree at the equator. We subtract 0.666 from the distance of 1° or 60 nautical miles at the equator and we have a distance of 59.334 nautical miles per degree at the latitude of 1° north. 59.33 x 360 = a cone circumference of 21,360.24 NM.

Extend beyond the exponential scale and we find:
1 Cosine 0.99984 x 900 = 899.86 Miles ÷ 15° = 59.99 x 360 = 21596 Circumference

Here is the table that shows the use of 0.0666 so that the measurers of the earth could get an idea of its nature and size.

Latitude in Degrees x distance in nautical miles = Distance around the cone or pyramid at that latitude.

```
0°  = 60 - 0     = 60     NM x 360 = 21,600.00 NM equal
1°  = 60 - 0.666 = 59.334 NM x 360 = 21,360.24 NM extended 21596
10° = 60 - 6.66  = 53.34  NM x 360 = 19,202.40 NM extended 21272
```

10° latitude equals a distance of 53.34 NM x 360 = a circumference around the internal pyramid or cone of 19,202.40 sea miles can easily be estimated and extended using cosines to 21272. Even today we use 60 nautical miles to measure distance in degrees from north to south or pole to pole and use that in combination with charts to find our position and measure journeys. The object of circumference measurement for the ancients was to measure distance as accurately as they could from east to west after measuring from pole to pole so that they could triangulate a course by the shortest route.

Now, it must be remembered that there is no evidence that equations were used in the time of the ancient Egyptians and yet David Ritchie has found considerable evidence in the size and proportions of the pyramid complex at Giza to show that they did. In fact he goes further to show a correlation of the 666 number between the Great Pyramid and the Sphinx.

A meridian is the name given to a line of longitude from pole to the equator and onward to the opposite pole. In modern times, the $0°$ meridian passes through the Greenwich Observatory in London, England, causing all local times in the world to be measured against the world time at Greenwich; this world time is called Greenwich meantime or Zulu. To find the difference of the distance of circumference at a parallel of latitude along any meridian of longitude in increments of minutes of arc or individual miles, divide the nautical miles by 60 minutes of arc.

On the imaginary pyramid, one second of arc is equal to 0.01666 or 0.01666 nautical miles. This last figure is hidden in modern navigation. Navigators these days use degrees and minutes of arc, but when they are taught the next division, they are instructed to use decimals instead of seconds and so the magic number does not show in the calculations. The other point that must be noted is that in modern geometry, measurements that are based on the polar meridians are earth based and that means meters. The more transcendental numbers of sacred geometry are based on the pentangle and the equatorial system of measurement is based in Egyptian cubits and that these measurements have passed down to the English cables, chains, yards, feet and inches. It seems that there is one number for the earth and a different one for God.[lix]

In the work by Erling Haagensen and Henry Lincoln Templars Secret Island[lx] more complex co incidences are found in units of measurement. For example, they suggest that the earth's circumference should be considered in feet as well as meters. And they quote:

"The equatorial circumference must be first divided into degrees, Each degree is then divided into as many parts as there are days in a thousand years.
1000(years) multiplied by 360(degrees) multiplied by 365.2 (days) = 131472000
NASA gives the equatorial measure as 400074156 meters, which is equal to 131476890 feet.
A discrepancy of less than one mile from absolute perfection."

They go on to explain various methods of sacred geometry as part of these ancient skills but we must stay with the basics. It has been said the pyramid could not exist without the cross since a method of surveying is essential before any construction, remember no archaeologist can effectively explain how it was done in any rational way. So until the present, it is a question that is shied away from in academic circles.

We know that the distance of a degree along the latitude of the equator is the same as a degree on the meridian of longitude, because they both measure the maximum circumference of the earth. However, the distance of a degree on a line of latitude decreases, as I have shown, as we move up the latitudes to the pole. The distance in nautical miles of one degree at ten degrees latitude is less than the 60 nautical miles at the equator and we find that difference in distance on the earth's surface by dividing the circumference at that latitude by 360°. 21272 divided by 360 = 59.08 nautical miles, therefore 1° of latitude at a latitude of 10° is equal to 59.08 nautical miles in distance east to west. One degree at a latitude of thirty degrees at Giza is equal to 18706 NM ÷ 360 = 51.96 nautical miles.

The number of the Beast for this type of measure of the earth in degrees, minutes and seconds, is also its fundamental measure of distance and speed. So, the cross and plumbline coupled with the pyramid shape of the ancient surveyors and astronomers exposes the number of the Beast as 666. All the observer had to do is sight the pole star and work out his latitude in degrees and minutes to find the pyramid shape within the earth at that latitude by using the 0.666 number as an incremental figure of division. The next necessary step for the observer to take was to draw a square within the circle of the earth and providing this square was set at 45° a pyramid form would be achieved.

The same pattern of the measuring rod on the cross and plumbline would be represented as the side of the pyramid forming an exponential scale using the number 666. Extending the measurements found through an exponential scale to the bow shape formed by the curve of the sphere would find the circumference at that latitude. Basic navigation would allow the observer to travel a measured distance up or down a meridian and then, turning east or west, use a known measurement translated into degrees along that parallel of latitude. By finding the angle to a star at the next latitude would allow triangulation to take place for the purpose of navigation.

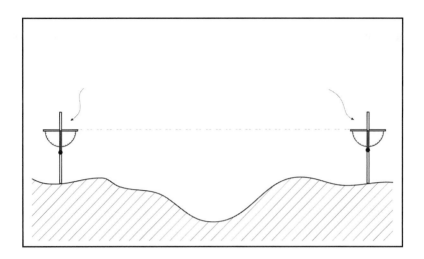

Surveying with the Cross and Plumbline

The skills of the builders of the Giza pyramids are admired and wondered at by all that study them. One of the difficulties facing the architect was that of building the pyramid on, and integrating it with, a rock outcrop or mound, let alone aligning it with true north. With this instrument, the planning of the pyramid would have been relatively straightforward, as it works like a modern-day theodolite. Also, by using this instrument, it is a relatively simple matter to incorporate angles of the outer and internal structures of the pyramid into the design. By first finding a level with the use of the plumb line and sighting mechanism, observations can be progressively taken using simple geometry to calculate the angles and distances involved with each observation.

If you know the distance and angle, the height can easily be calculated. Conversely, if the height is known, the distance can be computed. Gradients are determined by angling the sight of the instrument or cross, upward or downward as required. As the plumb line alters its position along the scale the angle can be read. A progressive series of readings can be taken over an undulating length of ground. A grid reference, east and west, north and south of the site can be built up showing rises and falls, ditches, gradients, hills, bumps and even mountains. The angles of slopes and gradients of distant mountains can be ascertained. These can be achieved by turning the instrument sideways and instead of sighting along the cross arm, using the staff of the Pesh-En-Kef to determine the angle or alternatively the length of the cross arm. It is only with a simple instrument like this, that the Pyramids at Giza could be integrated and aligned so perfectly with the topography.

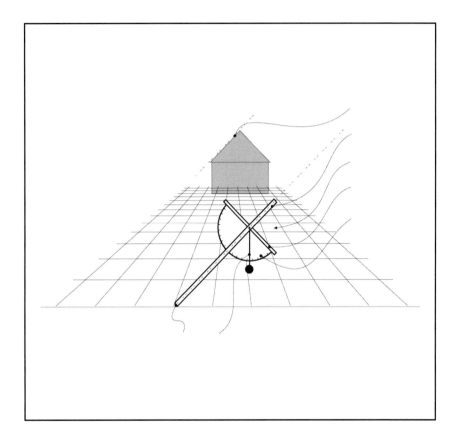

I showed the instrument to a friend called Moustafa Gaddalla at a meeting in Paris in 1998. Moustafa is a writer on ancient Egypt and author of several books including his work called "*Historical Deception, The Untold Story of Ancient Egypt*", he was a trained civil engineer from the University of Cairo. Moustafa had not heard of the instrument before and he said it would have been immensely useful in some of the road building projects that he had been involved with. It can be seen from the example above how an angle can be determined. The plumb line is always at 90° to the horizontal and the scale will compute the angle taken in a sidereal observation. This can be used for obtaining the angle of slope of both a near or distant object that already exists, such as a mountain or hill, building or roof pitch. It can also be used prior to the building process to determine the angle of a sloping construction or shaft, such as the shaft shown above right in the north face of the pyramid of Khufu at Giza.

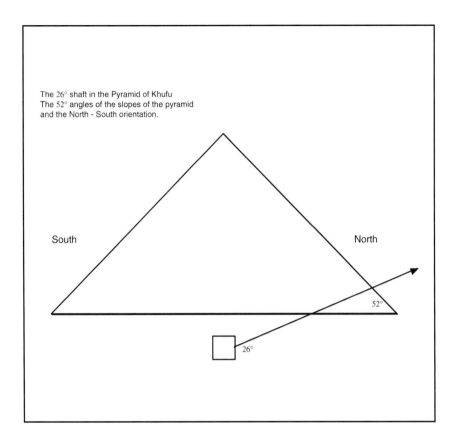

The 26° shaft in the Pyramid of Khufu
The 52° angles of the slopes of the pyramid
and the North - South orientation.

Having determined the angle from a point on the ground or foundation that the lower cross arm indicates to, a simple wooden rod or marker pole can be temporarily inserted in the place at the predetermined angle required. In the case of the 26° shaft sunk through the bedrock of Khufu's pyramid, this method could be employed successfully. The shaft is exactly half the angle of the exterior slopes of the pyramid.[lxi] The other shafts and the setting out of the proportions of the Queen's and King's chamber would be child's play compared with the systems used to survey the pyramid by Sir William Flanders Petrie.[lxii] They would not need spirit or water levels for this instrument, as the plumb line indicates level in all planes, in relation to its parallel attitude to the staff or upright and scale rule. In the same way that modern surveyors use an assistant with a measuring pole for working out gradients over distance, this instrument could be employed. This would help explain how the builders were able to incorporate the outcrop of bedrock into the structure of Khufu's Pyramid. It would also permit structures to be constructed on different gradients up to the Giza Plateau, overlooking the Nile. It is all very well to accept that a design may work on paper, but practical applications are quite different and only an instrument such as the one I propose, could carry this application through to material completion. Nothing of consequence is built without a thorough survey of the site first, both during and after construction.

Distance and Height

The distance from the observer to an object is simple geometry if the observer knows the height of the object from a chart or map. The instrument may be used for determining distance of a ship from a known object such as a charted mountain or lighthouse etc. The navigator must take into account his height from sea level and the distance to the horizon to correct any errors. If the height of the object is known, then the distance can be quickly worked out by the application of simple mathematics used by all navigators. The same is true for height if the observer knows the distance from the object.

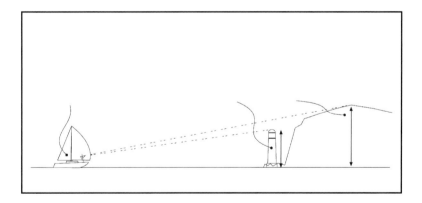

Solar Sights

As it is impossible to view the sun directly with the instrument, a method of operating the instrument to be able to achieve accurate solar positions would be as follows. The observer would stand at the sunward end of the sighting cross arm and project the shadow formed on the ground or other horizontal surface. If no lateral shadow of the instrument's arms are visible and the V sight or sighting tube projects the light of the sun to the centre of the shadow, then the angle is

187

correct and the degrees can be read from the scale. In modern sextants using mirrors, the geographical position of the sun is first found by bringing the image of the sun down until its edge touches the visible horizon. This is essential because the sun is an orb which depending on its position at the time of observation may cover a geographical area of up to 3000 miles. This problem of observing the centre of the sun is eliminated with the cross because of focus. If the observer reverses the instrument and projects the shadow image directly onto a backdrop, then the scale will automatically read the correct angle to the centre of the sun. Modern mariners are not taught about the ecliptic in astro navigation, what they are taught is how to reduce the angle read from the sextant by looking up the almanac for the day. In the case of the sun, it may vary in altitude in a given meridian position as much as 46.8° dependant on the time of year and that represents a potential error of up to 2,808 nautical miles.

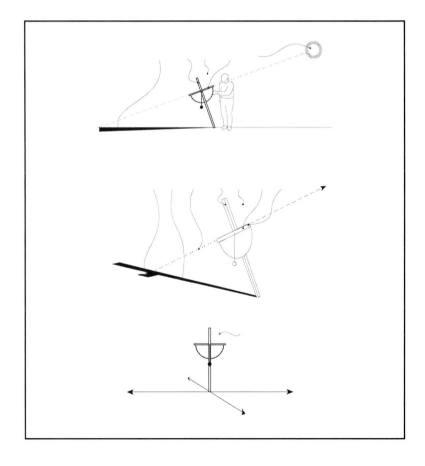

So, the conclusion is that sight reductions are not necessary with the cross and with the correct knowledge, neither is an almanac.

The ancient mariners were able to work out the predicted altitude of the sun above or below the equator with out the use of almanacs in the same way as any one can, as long as they mark the equinoxes and solstices in the calendar. It is also possible to reduce any position error further by taking further sights over a given time period and transposing the results in the following manner. If a mariner was using a small hand held instrument then the area of accuracy would be of 1° equal to 60 nautical miles. The first sight is transposed onto the chart as a circle measuring 60 miles. A later sight is also transposed and where the lines of the two circles intersect gives a good indication of position. This method is known to astro-navigators as "Sun run Sun". This can be seen in drawings from the Rennes le Chateau parchments, discovered by Berenger Saunier in 1891 where there are two overlaid circles divided by a pentagram which demonstrate navigation and surveying techniques through sacred geometry.

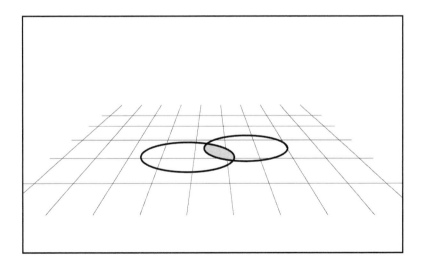

To Find Longitude

Since the invention of the clock and watch, we have taken time keeping for granted. However time keeping was very different in ancient times, it was a much more difficult exercise in navigation and only of benefit to mariners or those who travelled the vast deserts of the world. Time keeping was considered impossible before the invention of the marine chronometer, which was named after Chronos. According to Manfred Lurker in the *Dictionary of Gods and Goddesses, Devils and Demons*[lxiii], Chronos was the personification of time and "was often coincident in the late classical period with the figure of Aion'. He was often depicted as a bearded old man with a sickle and hourglass, especially in the Renaissance and Baroque periods.

The world is a spinning sphere and consequently it is impossible to take a sight on a star in the sky and find longitude without a vital ingredient. That ingredient being local time in relation to prime meridian time. In order for this to be calculated, you would first need to find latitude as already described and you would also need to know the time at 0° longitude. It has already been established that the original 0° longitude or prime meridian was found to have been at the Giza site. Mediterranean after all means "Middle of the Earth". The prime meridian was only moved to the Greenwich site in relatively recent times after great arguments between the English and French, who wanted the meridian to go through Paris.

The big quest in 17th century Europe was to invent a mechanism that could maintain accurate prime meridian time whilst voyaging at sea. This great enterprise was embarked on as a result of the large number of ships that were lost as a result of striking land due to a lack of being able to find longitude. These high levels of maritime losses resulted in the watch being invented and the method of this discovery has been well documented.

The device was so named after the job of watch keeping on board ship. The watch is based on a science designated as Horology named after Horus and the Watchers in ancient Egypt. To be able to locate longitude using this method requires that prime meridian time be kept on one timepiece. Another timepiece is adjusted to local time by sighting a star or the sun on the ecliptic and determining its position with a sextant and correlating that observation with the predicted position from an almanac. Having achieved this, the information can then be kept on a separate timepiece as local time and the distance measured from the prime meridian to the local meridian in addition or subtraction of degrees and minutes, which are adjusted to nautical miles.

For every 15° that one travels eastwards or westwards of a prime meridian, the local time moves one hour ahead or behind depending upon direction. Therefore, if we know the local times at two points on Earth, we can use the difference between them to calculate how far apart those places are in longitude, east or west. This idea was very important to sailors and navigators in the 17th Century. They could measure the local time, wherever they were, by observing the Sun, but navigation required that they also knew the time at some reference point such as Greenwich, in order to calculate their longitude. The actual distance in nautical miles is dependent on the latitude of the observer as the meridians draw closer in nautical miles, the further from the equator that the observer is located.[lxiv] One of the reasons that the latitude must be found first is so that the navigator can work out the distance horizontally at that latitude. The distances on a circumference of latitude are explained in the section on measurement.

In modern times, a chart usually based on a Mercator projection is essential for measuring distance and dividers are used to mark off the distance on the chart, using a measure of nautical miles taken from the divisions on the longitude scale.[lxv] A Gnomonic projection is used for higher latitudes where the meridians converge more dramatically. Prior to marine chronometers, sailors used hour or sandglasses and the duty of the Watch keeper was to turn these glasses and log the time of each turn. Most people are aware that a bell was rung every hour at sea to signify the turning of the glass, an example is that at 0600 hours, six bells are sounded. They also used a smaller, separate hourglass to record the distance travelled by measuring the speed of the boat through the water. This was achieved releasing a length of rope into the sea, with a drogue on the end, with knots tied at regular intervals. This rope was allowed to run for the period of the small sandglass and upon expiry of the sand, the rope was stopped, then recovered, counting the knots that had been expended, therefore enabling the speed through the water.[lxvi] The name knot is still used today to measure speed at sea.

However, this method has to be used as an aid in conjunction with other dead reckoning methods. The sea is a moving thing and the speed through the water is not always the same as the speed over the ground beneath. Many factors can affect this, not least being currents, leeway (the sideways drift of a vessel caused by lateral wind forces) and tacking manoeuvres. It is also believed that the Vikings, ancient Greeks and the Phoenicians used a form of magnetic compass in the form of Lodestones. Ancient mariners also used wave and wind direction as a guide.

Historians are already knowledgeable of the fact that the Egyptians traded with India and with Lebanon from where cedar was brought for the construction of their boats.*lxvii* It is also known that the Phoenicians traded great distances with many countries and it has been recorded that the Phoenicians circumnavigated Africa. Consequently, it is my belief that they must have used a system known as pilotage.

Pilotage is the method known and used by navigators for finding positions in coastal waters and requires the recognition of landmarks capable of being seen from the sea. By recognising, visually, certain mountains, bays, estuaries and townships etc., one can ascertain one's position whilst sailing along a coastline. The distance from the object can easily be measured using a known instrument called the cross staff which is used by making observations along its scaled length with a sliding cross piece, which will read off the distance. By using a plumb line for measuring depth with a knot system of measurement, underwater coastal gradients help with pilotage, particularly in the Mediterranean where the depth stays constant due to the lack of tides.

Charts are an essential requirement illustrating the topography and depth contours of the land. Tallow was often tied to the bottom of the lead or weight of the plumb line so that it would pick up sand from the bottom and the navigator then became aware of what sand or grit was native to the area. This was also an aid in navigation as the type of silt or sand could be analysed for pilotage. We know that they had charts from such finds as the Piri Reis map, which was faithfully copied generation after generation on calfskin.*lxviii*

We have reports from ancient historians and philosophers such as Plato and the Roman geographer Statius Sebosus, that there were islands to be found west of Africa and beyond the Pillars of Hercules. The Pillars of Hercules are known to be at Gibraltar on the western end of the Mediterranean Sea. The next islands to be found Southwest of Gibraltar are the Canaries, just off the coast of Northwest Africa. Several thousand miles further west and across the Atlantic is the location of the West Indies and the Gulf of Mexico. The trade winds blow at regular times of the year from the West Coast of Africa to the Gulf of Mexico in a circle around what used to be known as the Sargasso Sea. The north circle of the route arrives on the west coasts of Europe between Ireland and Spain. Ocean currents run from the Canaries to the Gulf of Mexico and the Caribbean Sea, small yachts sail the trade winds every autumn from the Canaries, following this same ocean current, to the Gulf of Mexico.

Modern private sailing boats as small as 17 feet are able to undertake this journey every November, in relative safety. Yet modern academia fail to acknowledge that the Phoenicians and Egyptians, who were capable of and did build sailing ships in cedar wood as long as 140 feet*lxix*, could have made this voyage with ease. So there is continued academic and historical stance, that despite their ability to build such craft, there is still no hard evidence that the ancients were capable of transatlantic voyages. This last historical stand is based on an argument that the ancient mariners were not capable of navigation and did not have the instrument capable of such navigation. This stubborn stand must now surely fall as I have shown that such an instrument was available to them and how it works.

It is often forgotten in our modern age of transportation of roads, automobiles and airlines that the safest and easiest way to travel and trade, up to the invention of railways, was by sea. Armies moved by sea and so did passengers and cargo, the countryside was full of forests, hiding robbers and bandits and was therefore not deemed to be safe. Foul weather caused mud and potholes for coaches and carts, which were extremely slow and cumbersome when compared with a sailing craft. In a fair wind, a sailing craft could carry hundreds of tons of cargo at an average speed of 5 knots, night and day, non-stop. That is a distance of 120 nautical miles every 24 hours, one thousand two hundred miles in ten days, three thousand six hundred miles in one lunar month, which was a much longer distance than the expanse between the Canary Islands and the West Indies.

Below is a list of the instruments, materials, knowledge and skills already proven to have been available to ancient mariners before my discovery of the far more sophisticated cross and plumbline:

1. Ships, high prowed and seaworthy.*lxx*
2. Sail power.
3. Oars.
4. Charts.
5. Crew.
6. Method of measuring distance travelled. Knots and nautical miles.
7. Cross staff for pilotage.
8. Astronomy.
9. A prime meridian.
10. Hour and sand glasses.
11. Watch keeping system.
12. An almanac.
13. The astrolabe and the cross-quadrant.

ASTROLABES

By kind permission of James E Morris, I have been permitted to quote his work on Astrolabes.[lxxi]

Firstly, what is an Astrolabe? James E Morris describes it as an *"instrument with a past and a future"*. It is a very ancient astronomical computer used for solving problems relating to time and the position of the Sun and the stars in the sky. Throughout the ages, several types of astrolabe have been made but by far the most popular type is the planispheric astrolabe, on which the celestial sphere is projected onto the plane of the equator. A typical old astrolabe was made of brass and was about 6 inches (15 cm) in diameter, although much larger and smaller ones were also made.

Astrolabes are used to show how the sky looks at a specific place at a given time. By drawing the sky on the face of the astrolabe and marking it so that certain positions in the sky are easy to find enables the user to find the information he is after. To use an astrolabe, you adjust the movable components to a specific date and time and once set, the entire sky, both visible and hidden, are represented on the face of the instrument. This allows a great many astronomical problems to be solved in a very visual way. Typical uses of the astrolabe include finding the time during the day or night, finding the time of a celestial event such as sunrise or sunset and as a handy reference of celestial positions.

Astrolabes became one of the basic astronomy education tools in the late Middle Ages. Old instruments were also used for astrological purposes. Though vital to seafarers, the typical astrolabe was not a navigational instrument but used to attain the altitude of the planets, although an instrument called the mariner's astrolabe was widely used at sea for navigational purposes. The mariner's astrolabe is simply a circular plate with a scale engraved around the edge and a bar with peepholes (the alidade) mounted on a central pivot. In use, the instrument was suspended from a pivot and the alidade was rotated until the star could be seen through the holes or until the shadow of the sun lay directly along the edge of the alidade. The altitude of the celestial body could then be calculated by reading the engraved scale.

The history of the astrolabe begins more than two thousand years ago. The principles of the astrolabe projection were known of before 150 BC but the true astrolabes were not considered to have been made until e.AD 400. It was an instrument that was highly developed in the Islamic world by AD 800 and was introduced to Europe from Islamic Spain (Andalusia) in the early 12th century.

It was the most popular astronomical instrument until circa 1650, when more specialised and accurate instruments replaced it. However, astrolabes are still appreciated for their unique capabilities and their value for astronomy. The largest astrolabe collection on public display is at the Museum of the History of Science, Oxford, United Kingdom.

The Mariner's Astrolabe as the image of the Celtic Cross.[lxxii]

The mariner's astrolabe was cut away in the form of the Celtic cross, with spaces in between the cross, to allow the wind to pass through to assist in stability.[lxxiii] An image of this type of astrolabe can be seen in a photograph of a stone frieze from the Yucatan. It shows a man in a canoe, with an astrolabe strapped to his waist. The photograph was taken by Mayan scholar Teobert Maler in the nineteenth century and kept by the American author Robert B Stacy Judd.[lxxiv] The first chart to embody cosmography information and show a quadrant, astrolabe, calendar with the zodiac and a declination scale, is the beautiful Ribero's World Chart of 1529.[lxxv] Diego Ribero was by birth, Portuguese. He became the royal cartographer in Spain and was responsible for revising and updating the official world map of the time (pardon real) and his map of 1529 is what he is most famous for. Because only two copies of the Ribeiro map are extant, a tracing of the Western Hemisphere portion was made by the nineteenth-century German historical geographer Johann George Kohl from the original copy in Weimar, Germany. The first map that indicated the true extent of the Pacific, it was the only 16th-century map that did not show China and Japan a short distance from the American West Coast.

Instruments Required for Finding Longitude

1. A Cross and plumbline (which is also a form of astrolabe), with the ability to take sidereal angular observations.[lxxvi]

2. A method of keeping and finding local time in accordance with the prime meridian.

3. A form of Almanac.

4. A Chart. The oldest surviving sea chart is the Pisan Chart, late 13th century made on vellum and using two networks of Rhumb lines, each forming a 16-point compass. The amazing observation when viewing this chart, is the accuracy. It is the most enigmatic of ancient charts or portolans and begs the question, how far advanced was chart making by this time or had this been a copy chart, made from knowledge stretching back for centuries earlier.[lxxvii]

The Solution for Finding Longitude

The instrument that was an important component towards the solving of the problem of longitude, is the method of using the instrument known as the cross and plumb line in the manner outlined earlier to find the voyager's latitude.

It is an obvious progression that the astrolabe developed from the quadrant. However, the astrolabe was too small to be accurate enough for navigation and in fact, the mariner's astrolabe used in the middle ages was inaccurate by as much as 5°. This is an unacceptable amount of inaccuracy and represents an error of 300 nautical miles, as I discovered earlier through my own experiments and can take you to a totally different destination from the one intended. A cross with a plumb line and at least 90 centimetres of scale rule gives an accuracy of between 3 and 6 nautical miles. A Celtic cross is far more stable at sea than the cross and plumbline and even one without a form of vernier scale would be accurate enough to find a position within the distance of the focal plane of the horizon.

By using the cross, the observer's latitude can be ascertained and the declination of planetary, solar, lunar, ecliptic and stellar objects can be measured. In so far as current knowledge has been established this method could only be used to find latitude. It is believed that the early navigators travelled along a line of latitude until they reached their destination. If wind and tide diverted their ship knocking them off course, they steered another course on a tangent (Rhumb line) until they picked up the correct or next latitude line again.*lxxviii* It must be remembered that to accept that they travelled along a line of latitude means that they were aware of the Earth as a sphere, as latitude may not be established without that knowledge. Latitude is based on the changing angle of a heavenly body as an observer travels up or down a meridian. However, to be more efficient in their voyaging, the determination of local time was the problem that our ancestors had to overcome. This was not an easy task for our predecessors. To assist them in being able to calculate local time, they, unlike our 17th century ancestors who were at war and in constant conflict, had to be at peace in the world, trading with other countries and kingdoms in mutual co-operation. It must be seen that they relied on co-operation and friendship of the peoples that they visited.

The Ancient Network of Horus

I believe that the ancients established a chain of star clocks or observatories all over the world to assist them in maintaining local time in relation to the gigantic and sophisticated clock at Giza, through which ran the prime meridian at that time. Using mirrors and beacons to advance the watch stations across the land, long forgotten surveyors originally developed the chain of ancient sites that we now know as stone circles, henges, pyramids and monoliths. This must have taken hundreds if not thousands of years to develop and calibrate, but calibrate it they did and each site was designated its own star that would be located over that site at the right altitude once a year, probably at the solstice.

Once the local star and local time was calibrated, it was only a matter of keeping records every night to ensure that the one degree drift of that star was noted and the time kept in relation to Giza which was the prime meridian. The most able astronomers of the time would have been given the task of manning these most important clocks and the scribes were given many vital functions. Functions such as the meticulous observations of all the stars visible in the Northern Hemisphere at night, the noting of the moon and its nodes, the observations of the planets and their trajectories. They interpreted the sun through knowledge of the sundial and its reflection not only on the Nile in the case of Akhetaten, but also through the use of conic reflection with clever foci on the polished surfaces of the pyramids and their strategically placed satellites.

I believe they measured the ecliptic and precession, the seasons and inundations, predicting times for sowing, reaping, voyaging and feasting. These times were kept with precision and the ancients lived in harmony with nature and the cosmos. They could not have achieved any of this without being able to measure angular differences both horizontally and vertically. I believe that the Egyptians inherited this knowledge from ancient mariners that chose to settle in the Nile delta long before the construction of the pyramids that we know. It must be remembered that there are Neolithic sites in the area similar to the ones in Scotland and Ireland and that these sites are far older than the pyramids. Is there any physical evidence that this is true and that such a network was created in prehistory? I believe that there is and I am not the only one. It is at this stage that I would like to introduce the remarkable work of David Ritchie, who has discovered such a system in the British Isles. This ancient system has many mathematical connections with the Giza complex as well as within the British Isles and David has already presented this evidence to Dr Zahi Hawass, The Egyptian Director of Antiquities. David delves further back into the history of Neolithic sacred geometry and mathematics of the islands of Great Britain and finds remarkable connections with ancient Egypt.

A SLICE OF PI*lxxix*

© 2001 by David Alan Ritchie

We divide the Earth's circumference into 360 degrees, with 60 minutes to the degree and 60 seconds to the minute. 360 degrees equals 21,600 minutes or 1,296,000 seconds. It has been that way since we inherited the system from the Babylonians over three thousand years ago.

It has never been questioned that it is the correct system for us to use when giving a location on the Earth a set of co-ordinates that can be found by navigation. As long as the Equator is zero latitude and Greenwich Observatory is zero longitude anyone with an accurate method of timekeeping and a means to calculate latitude can work out where they are standing and give it a number that can be located by someone else with the same tools. We take navigation for granted; yet I doubt that most of you have the necessary skills that would enable you to traverse the surface of the Earth unaided. You don't need to feel ashamed of that because in a modern world sailors and pilots only need those talents, the very people we depend upon to move us from continent to continent.

For most of us navigation is buying a road map and following the street signs until we reach our destination, but have you ever considered the problem in a historical context? Why did the Babylonians need to measure the circumference in 360 degrees, and how did they know the Earth was a sphere? We can make some assumptions as to where they acquired the knowledge of measuring the Earth, assuming, of course, that they didn't come up with the skills completely independently. Modern historians have tended to overlook the influence of Egypt and the knowledge they possessed of mathematics and measure, they were a culture who had been measuring the Earth and stars long before Babylon existed. It is easy to credit Babylon with the discovery of Earth measure, but first we should look at a little piece of history in a different light.

The Babylonian captivity of the Jews is familiar to most people, be they historians or not. It is well documented in the Old Testament and Jewish history, which also recounts the Exodus from Egypt with Moses. What few people realise is the reason why Nebuchadnezzar captured the Jews. He was searching for knowledge, because knowledge is power, and the Jews had knowledge that was greater than his own and he wanted it badly. He abducted King Zedekhaia and blinded him; he murdered his two sons and took the Jews to Babylon. The King's daughter, Teamhair, escaped, taking with her the Stone of the Covenant, but we shall return to her story later.

The first assumption to make is that since the Jews had been a part of Egyptian culture they would retain information that had been accumulated in Egypt, but the priests and nobility, not the masses, would hold it. Even though Egypt does not document the Jews or their Exodus, there are many historians who believe that Moses was a High Priest of Egypt and by consequence would have taken with him secrets that would have been taught to his successors. If this were the case, it is only logical to assume that it included the mathematical knowledge of Egypt, knowledge that, I believe, predated Egypt itself. The point I am making is that King Nebuchadnezzar was stealing the secrets of the Hebrews and the chances are that it included knowledge of the 360-degree system of measure, which we credit to Babylon without question, but I have a few questions of my own.

My research over the past few years has uncovered many secrets of how the Ancient Egyptians measured the Earth and the system of measure they utilised. Part of my discovery has been the geometrical system by which the Giza Pyramids are laid out and the mathematical equations and formulae they encode. We have much information about the fractional / decimal system they used found in the Rhind Mathematical Papyrus and the Moscow Papyrus, texts which few classical scholars go near, simply because mathematics is outside their area of study. In fact, of the hundreds of thousands of articles and treatises written on Ancient Egypt, fewer than ten have been written on the mathematical papyri. Historians do not base their translations and theories on what was a fundamental and essential aspect of Egyptian science. The Egyptians understood Pi.

The system of measuring the circumference of a circle in relationship to its radius and diameter was understood with precision by the Ancient Egyptians, as was the Pythagorean theory of right-angled triangles. Both are built into the Giza Pyramids. The Great Pyramid, Khufu's or Cheops Pyramid, has the proportions of seven high by eleven wide, therefore its height divided into its circumference (44 ÷ 7) produces the approximation of 2Pi of 6.285714... or 6 and 2/7ths. The second pyramid, Khafre's or Chephren's pyramid, encodes the proportions of the 3-4-5 triangle where its height equals the 4 side and its half base the 3 proportion and consequently its slope is the 5 proportion. These relationships are covered in greater detail in my book with Simon Cox, The Makers of Time, but I want to develop one area in particular in this appendix. It all has to do with the number seven.

Seven was the magical number in Egypt; their entire measuring system and mythology was based around it. The Giza Plateau geometries develop a system of geometry that produces a harmonious integration of the constructs of the pentagon, hexagon and heptagon, or 5, 6 and 7. Now we are getting close to the problem I have had to solve and I'll begin with the simplest demonstration of it I can find.

360 is a number that cannot be divided by seven and produce a whole number. 360 ÷ 7 equals 51.428571... nor can it be divided by eleven and produce a whole number, 360 ÷ 11 = 32.72727... 360 degrees is not based in Pi. It is a simple system of measure that works just fine and the Egyptians would have been well aware of it, but it wasn't Sacred. There is another number system that is far more harmonious and is based around seven, eleven, seventeen, fifty-one and fifty-six, that number is Pi of 3.1416 and 2Pi of 6.2832.

Modern precision engineering uses the same figures, a rounded off version of true Pi which is 3.141592654... On the size of the Earth, the difference between true Pi and 3.1416 is less than a hundred metres. So it makes little or no difference where accuracy of measure is concerned, and yet it is the basis of a system of measure that has been staring us in the face for thousands of years. All will be revealed shortly, but first another look at Babylon. If Nebuchadnezzar was stealing the knowledge of the Jews and, by default, the Egyptians, he was given only a part of the truth, a system based in 360 which was not the sacred measure based in Pi. Zedekhaia would not have been too happy at being blinded and his sons murdered, so is it possible that he was a little economical with the truth?

He could easily divulge a 360-degree system and Nebuchadnezzar would be happy with it because it works, but he obviously did not gain access to the sacred system based in 357 degrees. So far you probably think I'm making up all of this, so first let me explain how it works. Numbers are just numbers, you can use any system you choose to measure the Earth, use as many degrees in your personal system of measure as you like, just as long as you understand it and can make it work. We have been handed down a 360-degree system and have never questioned it because it is satisfactory and does the job. I want to show you my system and how it works before I tell you how I discovered it with the aid of the Giza geometry.

Divide the Earth into 62,832 lines of latitude and longitude; each division being a slice of Pi (pardon the pun). It is only another equivalent of minutes or seconds and we could easily make it 31,416 slices, but in this system each slice is a single unit of Pi or 2Pi. The Great Pyramid gives us the number clues for the next subdivision. As I pointed out earlier its proportions are 7 x 11. The height divided into two sides (22) gives an approximate Pi of 3 and 1/7th. (3.14285714) and into four sides (44) produces 2Pi. The Egyptians loved mnemonic numbers; ones that could be easily remembered and formed part of a much greater mathematical picture. They did not limit themselves to a singular system of measure as is seen in the fact that they used two values of Cubit, the Royal and the Sacred, but I will cover that in detail in my book. For now, let me show you how to slice the Pi.

31,416 divided by 22 creates a fraction, one that is fundamental to the entire system. That fraction is 1/1428th. 31,416 divided by 88 equals 1 /357th. That means simply that if we divide the Earth into 357 degrees each degree represents 88 units of Pi, and obviously 1/1428th equals 22 units of Pi or 44 of 2 Pi. It makes a little more sense when we look at the system of measure, which underlies this division, the Sacred Cubit. The Egyptians had a measure for the polar axis, the distance between the North and South Poles. It was 500 million inches, or 20 million Sacred Cubits of 25 inches. That becomes the diameter of the circle between the Poles, and that circle has a circumference of 20 million x 3.1416, or 62,832,000 Sacred Cubits. Do you see where this is going? One unit of 2 Pi equals 1,000 cubits on the surface of the Globe and one unit of Pi equals 2000 sacred cubits. Therefore, the 1/1428th fraction measures 44,000 Sacred Cubits. One side of the Great Pyramid measures 440 Royal Cubits, a different measure but the base number is the same. This is not the place to elaborate on the harmony of the Cubit system, but it is covered in detail in 'The Makers of Time'.

The Hebrew alphabet comprises twenty-two letters which each has a numerical value; the language itself forms words, which have numerical values. This system is ancient and is obviously based in mathematics; there is a hidden code within it. Jewish theologians have spent centuries trying to unravel the code that was implanted in their language from the beginning but the fundamental question has to be, why? The tribes of Israel, the Hebrew, came from Egypt where they were known as the Habiru, or foreigners. Prior to their Exodus they were not a cohesive culture, their ethnic roots were extremely diversified and Judaism was formed from their need to form a single theological-political structure. One tribe in particular, the tribe of Dan, did not adhere to the tenets of this new religion and left Israel before the Babylonian captivity and moved away to an unknown, by Hebrew history, destination. Many historians have speculated that they were in fact the Tuattha de Danann, a Celtic tribe based in Ireland. This speculation may not be unfounded, as you are about to see.

The Pharaoh in Egypt was surrounded and protected by a group of courtiers who were foreigners and were called the Iry-Pat. They took care of the everyday business of running Egypt because they were the equivalent of modern civil servants. It is quite possible that they were part of the Exodus, which would account for the event not being recorded in Egyptian history. It is also possible that they were the Tribe of Dan. This may seem like a huge supposition to make but as the evidence progresses you will begin to understand why I am taking this track and please remember this is merely an outline and I will not be presenting all the evidence in this article.

Early Egypt had a symbol, which was revered for thousands of years before it disappeared without comment, just like the Jews, it was the Ben-Ben stone. The hieroglyph for the Ben-been is a three-stepped pyramid, coincidentally the same as the small satellite pyramid of Menkaure's Pyramid, G3b, which is the key to solving the Giza geometry. The Ben-Ben was also known as the Stone of the Phoenix. It was kept atop a pillar at Helipolis in its own sanctuary. Little is known about the Ben-Ben, especially the significance of what it represented. One possible clue is its hieroglyph; if it is a true representation of the actual artefact then the chances are that it was 'The Measure', the yardstick that comprised the Sacred Measure of Egypt's founders. Measure is the fundamental of any civilisation; you cannot be taxed until you can be measured. Which brings us back to the lady whom escaped the Babylonian captivity, Teamhair, or Tara, daughter of Zedekhaia, the King who was economical with the Truth.

Legend says that Tara escaped with the prophet Jeremiah in 586 BCE, taking with them the 'Stone of the Covenant'. They travelled first to Egypt, then to Spain, where Tara married the King of All Ireland, Eochaidh, before moving to Ireland to become Queen. The Stone went with them and for a thousand years afterwards the Irish kings were crowned in its presence. Now if that short statement doesn't open up a thousand questions, you're not paying attention. First let's examine the Stone of the Covenant, or Jacob's Pillar as it was also known. We are told that the Jews left Egypt carrying the Ark of the Covenant, was the Stone the occupant of the Ark? I can surmise that the only covenant with God is Sacred Measure, the ability to measure His creation, and the most valuable artefact to a nation whose language is based in number had to be 'The Measure'. There are too many coincidences to explain, and the next is what Tara herself represented, the Bloodline.

Here is a refugee from a Middle Eastern country running away and marrying a King of Ireland, one of the Tuattha de Danann; it sounds like a fairy tale. And then the artefact she brings with her becomes the centrepiece of their coronations for a thousand years. I have to assume that these people had more in common than lust. They obviously shared a common belief and possible family ties, especially if the Tribe of Dan were the Tuattha de Danann and, if the Tribe of Dan were the Iry-Pat, the civil servants of the Pharaoh, then what I am about to reveal to you makes perfect logic. It's all about measure.

What role did the Iry-Pat play in Egypt and why did a group of foreigners have such influence? The Bible itself talks about a race of people who were Measurers, who carried their measuring staffs wherever they travelled. The iconography is familiar to anyone who has seen a picture of an Egyptian God or Pharaoh; they all carry a staff. The shadow of a staff of known length can determine your latitude on the surface of the Earth, let alone the possibilities of an instrument like Crichton's cross. Were the Iry-Pat a group of Celtic surveyors who ran the observatory that was the Giza Pyramids? Were they the true founders of Egypt? I can see the smile spreading across the faces of those of you who think I've backed myself into a corner, but this is where facts obliterate most of what we determine as true history.

Stonehenge and Avebury Circle fall on a line of longitude that is exactly 33 degrees west of the Giza Meridian. That's in a 360-degree system. However, that is the only part of the Grand Design I am about to show you that is in the 360-degree system. Giza is at latitude 30 degrees North, or 1/12th of a great circle north of the Equator. Stonehenge, at latitude 51°10.58'N, is exactly 1/17th of a great circle further north. Avebury, at latitude 51°25.71'N, is exactly 1/7th of a great circle north of the Equator. Pay attention at the back of the class! One twelfth plus one seventeenth subtracted from one-seventh leaves a fraction. That fraction is 1/1428th. The measure between Stonehenge and Avebury is 17 miles 633 yards, or 44,000 sacred cubits or 1,100,000 inches. It defines the 357 system.

This is merely the beginning of a process of discovery that has overwhelmed me for the last few years. It will change forever how we look at our history. If it merely stopped at these three locations I would admit to coincidence, but it doesn't. There is a grid system laid out across Britain and Ireland that is defined at its intersections by many famous places including Iron Age forts, stone circles, standing stones, castles, churches and cathedrals. The Romans built their roads along these lines of longitude, connecting different grid locations, and all of them triangulated to the Great Pyramid at Giza. I know it all sounds fantastic, but when you see it you will probably believe it, it goes far beyond coincidence and the odds against it being so are comparable to winning the lottery. Which leaves a problem, if I'm not mistaken. You see Stonehenge predates the Giza Pyramids by four or five thousand years, and yet it and Avebury are built in relationship to Giza. Now if you want to tell me that the Egyptians were here 10,000 years ago building wooden and stone henges, you will have to come up with some startling evidence. Otherwise it's the other way round; it was the Celts and Druids of Britain who were measuring time and the Earth.

Egyptian mythology holds a big clue to the truth. The God Set fought a battle with Horus, the son of Osiris. Set lost his testicles in the fight and was banished from Egypt, yet his city was still the destination of the Journey to Heaven in the "*Book of What is In the Am-Duat*" for souls departing to the Afterlife for thousands of years after his banishment. Horus, in the fight, lost his left eye but retained control of Egypt. Set was described as 'The Watcher in the North ' and 'Lord of the Northern Sky'. If Set was an astronomer from Britain who had returned from Egypt and the 'eye' that Horus lost was an observatory that covered Britain then this final, for now, clue may shine a light on why I think history is wrong.

There is another line of longitude that is well defined by ancient structures in Britain. It begins at Setley in Hampshire and as it travels north it passes through, in sequence, Grateley, Uffington Castle and the White Horse, the Rollright Stone Circle, Meriden, Kirk Langley, Carl Wark, Tingley, Kirkstall Abbey, Ripley, Fountains Abbey, the Thornborough Stones. Darlington, Durham Cathedral, Chester-le-Street, the eastern end of Hadrian's Wall at Wallsend. Amble Castle and then runs into the North Sea at Dunstanburgh Castle. It then passes through Fair Isle before it reaches Shetland and terminates at the Giant Stones of Hamnavoe. It began at Setley and ended at Zetland or Setland. That well marked line with so many towns whose name ends in Ley and so many famous structures, is exactly one eleventh of the circumference of the Earth west of Giza. Khafre's pyramid, in the Giza geometry, defines that precise division.

This has only been the briefest of introductions into the enormous amount of evidence that Simon Cox and myself have accumulated. I will close with one other enigma. If 360 degrees can be questioned so can 24 hours in the day. If it was 22 hours in the day, 51 minutes to the hour and 56 seconds to the minute then the Eleventh Division would be exactly 2 hours west of Giza and there would be 1122 minutes or 62,832 seconds to the day. I hope you are intrigued enough to examine the final work which is still being written at this moment, but I can promise you an insight that nobody is prepared for. There is much, much more to come, but that's Pi for now folks.

© 2001 David Alan Ritchie.

David Ritchie's work is extensive and it is very hard to find room for coincidence in his enormous discovery. The finding of latitude and longitude by the ancients coupled with the precise siting of their monuments on a grid system that David has discovered is very hard to refute because there is an enormous amount of physical evidence to support him. David and I first got together when he heard from Simon Cox that I had discovered an ancient instrument that could carry out such measurements. David talks of latitude in his work and I say again, it is only possible to find latitude on a sphere, therefore the surveyors of antiquity knew the Earth was round.

Furthermore, that knowledge requires the understanding of spherical geometry by the observer and the only way to find coordinates without a satellite navigation system is to measure the angles of a heavenly body in relation to the Earth's centre. To achieve this requires an inclinometer and that is exactly what the cross and plumbline or Celtic cross is. Again we have evidence of the cross being as ancient as the giant stones of Hamnavoe.

It has long been suspected that most churches in Britain were built on much earlier prehistoric sites and there is much speculation that Ley lines connected these sites. I have already suggested that the term Ley line has been taken from the navigational term known as lay line, which is a reference to the method used by sailors to lay a course to a destination. It is not right to reject concepts out of hand and it might be true that the ancients had some kind of knowledge of the magnetic lines of the Earth. It is through reason and common sense that I believe all of this knowledge that I am revealing to you is nautical in origin and Ley line is a nautical terminology that is ancient but still used today. The term "*as in heaven so below*" is a great clue to the methods of astronomy used by the ancients to determine local time and a considerable support to my work.

A further supportive clue is given in the New Testament in the story of the three wise men from the east who followed a star to the birth of Jesus. The ancient mariners and the travellers on the desert followed stars and they achieved this by observing what star was at the meridian at midnight on a predicted day. They compared this star to the distance in degrees and minutes to the star that was over the place where they wished to arrive, triangulated its position against the pole star and followed it to the destination. If a star is accounted as being over the ancient prime meridian at Giza at midnight, then all other stars can be allocated a position over other meridians of longitude and parallels of latitude.

As I have already said, building this system must have taken thousands of years and I believe that David Ritchie is right and that the builders of the great stone henges reconstructed this system of navigation, surveying and time keeping. These builders are proposed as the Neolithic peoples, however, we must take into consideration the dramatic effects of the flood and its destruction upon the world. We are forced to look further back to Palaeolithic times and the evidence of knowledge of the zodiac in French cave paintings and the Clovis points on arrowheads found on both sides of the Atlantic that employed the same technology. In conclusion, I believe that Neolithic peoples constructed stone circles, water clocks and pyramids along the coastline at places they were voyaging to and from as staging posts. They were constructed as means of being able to maintain local time in relation to the meridian time at the site of the Great Pyramid at Giza. I believe they employed watch keepers from whom they obtained almanacs, charts and local time to be compared with Giza, the world clock. These watch keepers would be trained to a very high standard, through a programme of apprenticeship or initiation and would be the keepers of the religious and agricultural timetables on behalf of the local populace, whilst being able to update the mariners if and when required. They would maintain their secrets and mystique, to protect their income from outsiders. Evidence of this is still seen today in the rites of secrecy sworn by the Masons.[lxxx] It is only recently that the Trade Guilds have been disbanded.

The Phoenicians were the masters of the seas in ancient times and as a consequence, they kept their trade secrets deeply hidden from friend and foe alike.[lxxxi] These master mariners were able to rebuild an ancient grid reference of the Mediterranean and the Indian Ocean and gradually, over a period of time, the coast of Europe and Britain would be visited. The ancient henges and monoliths were resurrected and added to their grid references for local time keeping, thus enabling further distances to be traversed and far off oceans to be navigated. These 'staging posts' that were old before the Egyptians and Phoenicians, are still to be seen today along the coasts of Cyprus, Malta, the Balearics, Spain, France, Cornwall, England, Wales, Scotland and Ireland. There are some still to be seen in the Baltic Sea, Scandinavia and Norway. These stone and wooden circles, pillars and chalk carvings were all clocks for keeping local time in relation to Egypt. They are capable of predicting climatic and seasonal changes, indicating times to plant, reap and sow, and even when to marry and have children. The ancient fertility festival of Mayday in England is timed perfectly so that the majority of mothers would give birth in January when no work is done in the fields.

Stonehenge was the central clock for ancient England. Callanish is probably the one for ancient Scotland. Their keepers or Watchers would have inherited untold powers over the population. The meridians and grid lines between these known observatories may have given rise to the ancient myth of 'Ley lines' which as I have already said, derives from the nautical term 'lay line' used for sailing ships to aid in navigation in combination with tacking manoeuvres.*lxxxii* Many of the sighting stations in Britain had churches built on them with the expansion of Christianity. The Spanish Conquistadors in South America were seen to have carried out the same practice. It is probable that this increased stellar knowledge of the ancients led to an enhanced awareness of man's place in the cosmos and the cycles of the seasons, life, death and rebirth. This knowledge may well have led to the creation of religion, faith, stability and justice. The Watchers were well aware that once a year, the stars above were directly over fixed points on the Earth's surface and that angles could be observed between these points with the use of spherical geometry, identifying places on Earth with these stars and constellations. So how did they travel at sea and cross oceans?

AN OCEAN CROSSING

The whole system of navigation was dependent upon the knowledge of the traveller that the world was a sphere and the very construction of Khufu's Pyramid and the cross, conveys this understanding.

Solstices and eequinoxes were, and still are, crucial times for mariners. Equinoxes bring gales as temperatures change in the Gulf of Mexico creating violent low-pressure systems that can sink ships at sea. There were and are times for sailing in areas of the world's oceans. At the time of the Fourth Dynasty in Egypt, Orion, at its zenith in the midnight sky, heralded the Autumn Equinox. This meant that it was the time for a transatlantic crossing.

The last outposts of the African and European continent are the Canary Islands, which, as already mentioned, are also home to a series of Pyramids constructed from volcanic rock. Their alignment is east to west on the islands of Tenerife and Las Palmas. Close to the most westerly pyramid on Las Palmas, a carving of what appears to be a Mayan face has been discovered. Many stone spirals and other carvings are on display in the Museum on Tenerife.*lxxxiii* So are mummies of the Guanche.

What then could have been the method of navigation to allow a crossing of the

Atlantic Ocean? As I have previously outlined, a method of watch stations would be established keeping time on a local basis, using the time difference between the stars at the prime meridian at Giza and relayed along the meridians westward at fairly regular intervals, topography allowing. This system of time-keeping could have been spread northwards as well as south along the African Atlantic coast to Mauritania and beyond to all worthwhile navigable and trading areas. Senegal was the ideal embarkation point to reach Brazil and the Caribbean.

This transatlantic crossing would be the most difficult of navigational voyages were it not for the seasonal trade winds; the first crew to have successfully navigated the crossing would have been very courageous, not knowing what exactly was out there to encounter on their travels. Navigating the coasts of the Mediterranean would be a relatively easy task as the distance to be negotiated between the watch stations would be comparatively short. Since a journey of 600 miles equals 10°, a watch system with sand glasses could easily manage that with fair accuracy. However the voyage to the West Indies from the final landmass before the vast expanse of ocean, the Cape Verde Islands is approximately 2,500 miles. The Cape Verde Islands are 385 miles off the coast of Mauritania and are the closest port to the West Indies and are still classed as an important communication station today, being a crucial location for both air and sea refuelling.

It is reported that Christopher Columbus took only 33 days to reach the Antilles in the Caribbean. His boat was considered to be a crude and bulky affair when compared with the ships of the Phoenicians which were designed with their sleek lines, high stems and sterns. They were designed to cut through waves easily and coupled with a highly efficient rig and sail plan, they would have had a better windward performance than the square rig that Christopher Columbus used in his more cumbersome ship. Although the Atlantic crossing is mostly down wind on the Easterly trade winds, a modern 45-foot yacht can make the same passage in as little as 15 to 18 days. The Phoenicians' progress was governed by their waterline and their maximum hull speed was approximately 7 knots. The Egyptian funerary ship of Khufu was discovered buried in the boat pits alongside the pyramid in Giza. It is said to be the world's oldest boat and was found stored in 1,200 pieces, ready for assembly. The boat was discovered alongside hieratic signs, indicating how it should be constructed, and a bit like a modern day flat-pack kit. Although it has been generally accepted that this ship was not used for ocean travel, it clearly demonstrates the technical expertise of the shipbuilders of the time.[lxxxiv]

If a Phoenician ship averaged a speed of 7 knots, it can be calculated that it would cross the 2,400-mile passage along the latitude line of 17° from the Cape Verde Islands to Antigua in just 11 days.[lxxxv] The distance in degrees is 40° of longitude before they would make landfall.

To carry out the transatlantic crossing, the star at the meridian on local time could be used for observation just before embarking, and the watch throughout the journey would maintain the local time at Cape Verde. Calculations for the drift westward of that star of approximately 1° every night, would be recorded by the watch keeper on the ship. The position of the star at midnight Cape Verde local time would then be adjusted against its observed position from the navigator's point of view. The difference in degrees between where it actually is and where it should be at Cape Verde would then be subtracted and converted into miles at that latitude.

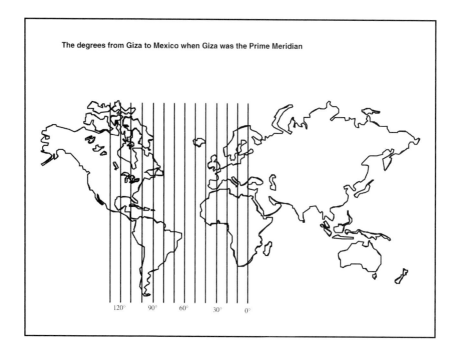

The degrees from Giza to Mexico when Giza was the Prime Meridian

120°   90°   60°   30°   0°

It was also possible to calculate local time through observation of the time elapsed which was recorded by the hourglasses and the marking of the star at midnight according to Cape Verde time. As the ship travels west and the star, which should be over Cape Verde at Midnight, was observed to be at an angle of 60° instead of 90°. When coupled with an adjustment of 1° for every day travelled, the local time could be calculated to be 22.00 hours. On arrival at the final destination, the navigator would head immediately for the local observatory pyramid or stone circle. A "Watcher" would be in residence and together with his priestly assistants, would confirm the local time in relation to Giza and all the watchtowers visited in between. A star viewed on the longitude of an observer in the Lesser Antilles was the star above the Giza meridian six hours earlier. That same star was on the longitude of the Canaries three and half-hours earlier and two and a half hours after being on the Giza meridian. The first leg of any transatlantic voyage made by the ancients, would be from Las Palmas (which calculated on the Greenwich prime meridian has the following coordinates: Latitude 28° 37' N Longitude 17° 56' W). Las Palmas is the most westerly of the Canary Islands to the Cape Verde Islands. It is located at Latitude 17° 02' N Longitude 25° 16W.*lxxxvi* To use the ancient system, add 30° to the longitude, thus allowing for the 0° longitude, or the prime meridian, to be located at Giza.

To reach the Cape Verde Islands from the Canaries, a distance of 900 miles, without following the African coastline, would require an incredibly accurate piece of navigation. It would necessitate the navigator to travel at a tangent of 8° 08' west and 10° 47' south. A slight error would result in the next landfall being the Weddel Sea in Antarctica. However, it is believed that the first voyagers would not have been able to undertake the navigation so precisely and so it is more likely that the sailors left Senegal at 17° latitude bound for the Cape Verde Islands.

It has been rumoured that seafarers from Senegal did in fact cross the Atlantic in the distant past. Furthermore, the shortest route across the Atlantic is from the Cape Verde Islands to Fortaleza in Brazil, a distance of only 1,693 miles. This was a point that was found not to have been missed by the Portuguese when negotiating their agreement with Spain over the division of the New World, the Portuguese were permitted to claim Brazil as their own. Was this intercession with the Spanish just luck, or were the Portuguese already aware of the existence of Brazil and more importantly, were they aware of its proximity to their homeland? It is without doubt, that the local Senegal sailors would certainly have been aware of the Cape Verde Islands and it is likely that the Phoenicians would have learned of their whereabouts from the locals during their trading excursions. Having established over a period of time the co-ordinates of the Cape Verdes in relation to The Canaries, a more direct route could be established between the two groups of islands. However, this would have required the knowledge of longitude and it is my proposal, that the Phoenicians not only knew of longitude, but were also capable of calculating it.

There is really only one ideal time to set sail for the Caribbean from Europe / Africa in full knowledge that there will be the assistance by the trade winds, westward around the southern side of the Sargasso Sea. This time is the period of late autumn / early winter. This is still true today and yachts leave the Canaries in late November / early December with a view to arriving in the Caribbean in time for Christmas and the winter season. They first sail Southwest to pick up the trade winds and navigate their crossing from that point, taking advantage of the winds. Ancient navigators could follow the same method providing they could read latitude to within a few minutes of arc.

I propose that this route is exactly 17° of latitude and runs from the Traza region of Mauritania, directly out to Porto Novo on Santo Antao, the outermost island of the Cape Verde islands and on across the Atlantic to Nelson's Dockyard in Antigua. This landfall is 43.5° west of the Canaries at Latitude 17° 0' N Longitude 61° 43' West. The problem arises in knowing when landfall was due to be made. A mistake in the calculations and estimations could easily result in shipwreck if it was made at night, in a storm or poor visibility. Many skills were available to ancient mariners, some of which I have already highlighted. To assist them in their navigation, they also used birds as a guide for the reason that many types of sea bird fly home to roost at night and seldom travel far from land when fishing at sea. These little creatures would act as an excellent guide for the seafarers when approaching land. Cloud formations are also an indicator of landfall. Wave formations are a sign of mountains up wind and low lying clouds top islands in the morning as humidity is carried aloft on the hills, condensing against the cooler air higher up. All of these things must have aided the first mariners, but the greatest asset to all mariners, without a doubt, is the knowledge of longitude.

ORION - THE TRANSATLANTIC NAVIGATOR

Due to the effects of precession, the seasons were found to have moved forward one day or one degree every 72 years. As a consequence of this, the star signs also change gradually. Orion, which was of great importance to ancient navigators, can now be seen at its highest point in the sky due south, at the winter solstice.

According to some archaeologists and Egyptologists, the year of 2400 BC was approximately the era of the construction of the Pyramid of Khufu. 2,400 BC was long after the time of the megalithic building phase in Northern Europe and Orion could be seen at its highest point in the south at midnight, on or around the 21st of October with Taurus extending to the west. Al Nitak (the lowest of the three stars) on the belt of Orion marks the position of the celestial equator. This was the period of autumn in the days of the ancients, close to the equinox, and the time when the trade winds began blowing westerly over the Atlantic Ocean. The Pleaides or Seven Sisters are 35° west of Betelgeuse when viewed from north of the ecliptic.

In the autumn, the Pleaides would be at a declination of approximately 17° due south at 22.00 hours. 17° latitude is 10.49° less than the latitude of Las Palmas at the Canaries and translated onto the meridian, this means that it is exactly 629.4 nautical miles south which represents approximately 4 days of sailing. A simple method of navigation would be to stay on a course due south, using the pole star to check the ship's latitude every night at midnight until the Pleaides are directly overhead, allowing the 1° shift westward of the stars every 24 hours. It is for this reason, that I believe that the Pleaides were so important to the Mayans as well as other ancient peoples. Having reached latitude of 17°, the ship would turn west and follow the latitude line. So, Orion the Hunter is the hour hand of the annual and precessional clock and was of great importance to the ancients. This constellation was known to many as the thirteenth sign of the zodiac and was represented by the ancient Egyptians as Osiris. The New Testament of the Bible also shows Orion and the zodiac in the image of Christ and his twelve disciples, with many references to the zodiac signs in the story, such as Fishers of Men (Pisces) and Shepherd of the lost sheep (Aries or Amen). The legend of King Arthur and the knights of the Round Table also bear a resemblance to Orion and the twelve signs of the zodiac. Clearly, Orion had a significance that was so great that, according to Robert Bauval and Adrian Gilbert in *The Orion Mystery*, the House of Ram built the three pyramids at Giza as a representation in stone of the three stars of Orion's belt upon the Earth. This importance placed upon the image of Orion is extended beyond the African Continent to the American Continent. But before we visit the image of Orion on the Earth in America, there is an image of this constellation in England, which I wish to make you aware of.

THE LONG MAN OF WILMINGTON

The Long Man of Wilmington is located near Brighton in the county of West Sussex, England. It is one of the most impressive mysteries in Britain. It is the largest representation of the human figure in Western Europe. It is 230 feet long and its origins and purpose are unknown to historians. Its location is N50° 49' 00" by E 000° 11' 331" and it is at an altitude of 207 feet above sea level. Carved in the chalk of the downs, it faces north so that an observer facing south at night would see its image in the sky above. That image would be Orion. Above it on the horizon can be seen cairns and burial mounds and it is almost certainly Neolithic in origin since these are dated between the bronze age at 1000 BC and the new stone age at 3000 BC. The site is located beside flint mines dating as far back as 3500 BC. It is also located beside the English Channel, on a headland, which makes it a convenient observation point for mariners and farmers alike. The earliest known reference is in the 18th Century Burrel Manuscript. Aerial photographs show that the Longman is elongated, but when seen from the ground, he assumes normal proportions. There are similarities seen in Roman coins and a 7th century gilt bronze buckle found in Kent.

Sophisticated surveying and astronomy was used to place this image of Orion on the hillside and this could only have been achieved with the cross and plumbline at a date earlier than the construction of the Pyramids at Giza.

THE WORLD'S MOST ANCIENT SYMBOL?

This deodogram is known as Nwit and means walled city, it is said to represent a round settlement and is documented as one of the oldest hieroglyphs. In most forms it can be seen with its axis at 45° to the perpendicular.

Does this represent the spring equinox or Orion's belt? The same offset angle can be seen at the three pyramids at Giza as pointed out by Robert Bauval and Adrian Gilbert in the Orion Mystery. Versions of the cross can be seen not only in Egypt and the Middle East, but also in the glyphs, friezes and pottery of the Amerindians across the Atlantic Ocean.[lxxxvii] A point to notice here is that this angle is also the layout for the majority of stone circles. Stone circles such as Stonehenge and Templewood in Scotland have a similar offset design to the Nwit and in my opinion, this demonstrates a worldwide astronomical knowledge and also indicates that the most important event of the calendar year was the spring equinox when time renews itself.

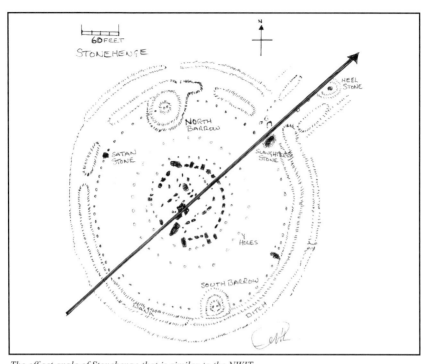

The off set angle of Stonehenge that is similar to the NWIT.

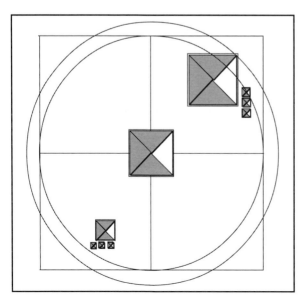

Here again we see the pattern repeated in the three pyramids of Giza, there is a representation of the similar angle displayed in the much older observatory of Stonehenge and the angle of the NWIT. Is this evidence that the glyph known as the NWIT meant observatory and not round village or walled city as previously thought?

All these constructions are also similar to the shape of a Celtic cross and certainly required such an instrument to survey them. A greater part of the evidence for transatlantic voyaging, astronomy and surveying in combination with the thirteenth constellation of zodiac. Representations of Orion can be seen across the Atlantic in America.

I will now introduce you to the work of respected researcher Gary David, and the discoveries he has made among the Hopi Indians in Arizona. The Hopi Indians are known as a gentle people and the most humble of the Amerindian Tribes according to their legends. Their history tells them that they emerged from underground caves after the deluge and took the least from God in their modesty, promising to keep the faith and knowledge until the next cycle of time. The Hopi were to encompass the stars and Orion in to their culture, religion and history in a remarkable way that is similar to Egypt.

This can be clearly demonstrated in the following article by Gary David.

THE ORION ZONE: ANCIENT STAR CITIES OF THE AMERICAN SOUTHWEST

by
Gary A. David

A Stellar Configuration on the High Desert

To watch Orion ascend from the eastern horizon and assume its dominant winter position at the meridian is a wondrous spectacle. Even more so, it is a startling epiphany to see this constellation rise out of the red dust of the high desert as a stellar configuration of Anasazi cities built from the mid-eleventh to the end of the thirteenth century. In fact, Orion provided the template by which the Anasazi (ancestral Pueblo people) determined the locations of their villages during a migration period that lasted centuries. Spiritually mandated by a god the Hopi call Masau'u, this "terrestrial Orion" closely mirrors its celestial counterpart, with prehistoric "cities" corresponding to *every* major star in the constellation. The sky looks downward to find its image made manifest in the Earth; the Earth gazes upward, reflecting upon the unification of terrestrial and celestial.

Extending from the giant hand of Arizona's Black Mesa that juts down from the Northeast, three great fingers of rock beckon. They are the three Hopi Mesas, isolated upon this desolate but starkly beautiful landscape to which the Ancient Ones so long ago were led. Directing our attention to this "Centre of the World," we clearly see the close correlation to Orion's Belt. Mintaka, a double star and the first of the trinity to peek over the eastern horizon as the constellation rises, corresponds to Oraibi and Hotevilla on Third (West) Mesa. The former village is considered the oldest continuously inhabited community on the continent, founded in the early twelfth century. As recently as the early twentieth century, the construction of the latter village proved to be a prophetic, albeit traumatic event in Hopi history. About seven miles to the east, located at the base of Second (Middle) Mesa, Old Shungopovi (initially known as Masipa) was reputedly the first to be established after the Bear Clan migrated into the region circa AD 1100. Its celestial correlative is Alnilam, the middle star of the Belt. About seven miles farther east on First (East) Mesa, the adjacent villages of Walpi, Sichomovi, and Hano (Tewa) - the first of which was established prior AD 1300 - correspond to the triple star Alnitak, rising last of the three stars of the Belt.

Nearly due north of Oraibi at a distance of just over fifty-six miles is Betatakin ruin in Tsegi Canyon, while about four miles beyond is Kiet Siel ruin. Located in Navaho National Monument, both of these spectacular cliff dwellings were built during the mid-thirteenth century. Their sidereal counterpart is the double star Rigel, the left foot or knee of Orion (we are conceptualising Orion as viewed from the front). Due south of Oraibi approximately fifty-six miles is Homol'ovi Ruins State Park, a group of four Anasazi ruins constructed between the mid-thirteenth and early fourteenth centuries. These represent the irregularly variable star Betelgeuse, the right shoulder of Orion. Almost forty-seven miles Southwest of Oraibi is the primary Sinagua ruin at Wupatki National Monument, surrounded by a few smaller ruins ("Sinagua" is the archaeological term for a group culturally similar and contemporaneous to the Anasazi). Built in the early twelfth century, their celestial counterpart is Bellatrix, a slightly variable star forming the left shoulder of Orion. About fifty miles Northeast of Walpi is the mouth of Canyon de Chelly, where another national monument is located. In this and its side Canyon del Muerto a number of Anasazi ruins dating from the mid-eleventh century are found. Saiph, the triple star forming the right foot or knee of Orion, corresponds to these ruins, primarily White House, Antelope House, and Mummy Cave. Extending Northwest from Wupatki/Bellatrix, Orion's left arm holds a shield over numerous smaller ruins in Grand Canyon National Park, including Tusayan near Desert View on the south rim. Extending southward from Homol'ovi/Betelgeuse, Orion's right arm holds a nodule club above his head. This club stretches across the Mogollon Rim and down to other Sinagua ruins in the Verde Valley. As a small triangle formed by Meissa at its apex and by Phi1 and Phi2 Orionis at its base, the head of Orion correlates to the Sinagua ruins at Walnut Canyon National Monument together with a few smaller ruins in the immediate region.

If we conceptualise Orion not as a rectangle but as a polygon of seven sides, more specifically an "hourglass" (connoting Chronos) appended to another triangle whose base rests upon the constellation's shoulders, the relative proportions of the terrestrial Orion coincide with amazing accuracy. The apparent distances between the stars as we see them in the constellation (as opposed to actual light-year distances) and the distances between these major Hopi village or Anasazi/Sinagua ruin sites are close enough to suggest that something more than mere coincidence is at work here. For instance, four of the sides of the heptagon (A. Betatakin to Oraibi, B. Oraibi to Wupatki, C. Wupatki to Walnut Canyon, and F. Walpi to Canyon de Chelly) are exactly proportional. While the remaining three sides (D. Walnut Canyon to Homol'ovi, E. Homol'ovi to Walpi, and G. Canyon de Chelly back to Betatakin) are slightly stretched in relation to the constellation— from ten miles in the case of D. and E. to twelve miles in the case of G (see diagram).

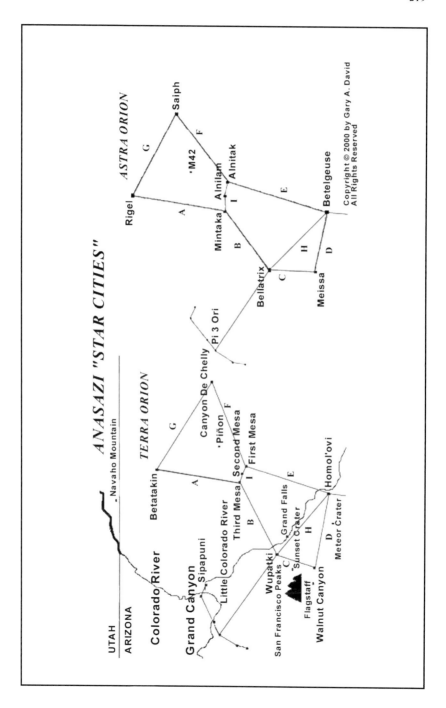

ANASAZI "STAR CITIES"

This variation could be due either to cartographic distortions of the contemporary sky chart in relation to the geographic map or to ancient misperceptions of the proportions of the constellation vis-à-vis the landscape. Given the physical exigencies for building a village, such as springs or rivers, which are not prevalent in the desert anyway, this is a striking correlation, despite these small anomalies in the overall pattern. As John Grigsby says in his discussion of the relationship between the temples of Angkor in Cambodia and the constellation Draco, "If this is a fluke then it's an amazing one.... There is allowance for human error in the transference of the constellation on to a map, and then the transference of the fallible map on to a difficult terrain over hundreds of square kilometres with no method of checking the progress of the site from the air."[1] In this case we are dealing not with Hindu/Buddhist temples but with multiple "star cities" sometimes separated from each other by more than fifty miles. Furthermore, we have suggested that the "map" is actually represented on a number of stone tablets given to the Hopi at the beginning of their migrations, and that this geodetic configuration was influenced or even specifically determined by a divine presence, viz., Masau'u, god of Earth and death.

Referring once more to the Diagram, we also note the angular correspondences of Orion-on-the-Earth to Orion-in-the-sky. Here again the visual reciprocity is startling enough to make one doubt that pure coincidence is responsible. Using Bersoft Image Measurement 1.0 software, however, we can correlate in degrees the precise angles of this pair of digital images seen in the Diagram. (Note: The celestial image is drawn from Skyglobe 2.04.)

Angle	Degrees	Difference
AG terra	65.37	
AG Orion	71.19	5.82
BC terra	132.60	
BC Orion	130.77	1.83
CD terra	84.31	
CD Orion	100.07	15.76
DE terra	97.79	
DE Orion	95.65	2.14
FG terra	56.17	
FG Orion	64.23	8.06

The closest correlation is between the left and right shoulders (BC and DE respectively) of the terrestrial and celestial Orions, with only about two degrees difference between the two pairs of angles. In addition, the left and right legs (AG and FG respectively) are within the limits of recognisable correspondence, with approximately six to eight degrees difference. The only angles that vary considerably are those that represent Orion's head (CD), with over fifteen degrees difference between terrafirma and the firmament. Given the whole polygonal configuration, however, this discrepancy is not enough to rule out a generally close correspondence between Orion Above and Orion Below.

Egyptian Parallels to the Arizona Orion

In their bestseller *The Orion Mystery*, Robert Bauval and Adrian Gilbert have propounded what is known as the Star Correlation Theory[2] (their book, incidentally, provided the initial impetus for writing the present article and book-in-progress). These co-authors have discovered an ancient "unified ground plan" in which the pyramids at Giza form the pattern of Orion's Belt. According to the configuration described very briefly here, the Great Pyramid (Khufu) represents Alnitak, the middle pyramid (Khafra) represents Alnilam, and the slightly offset smaller pyramid (Menkaura) represents Mintaka. In addition, two ruined pyramids - one at Abu Ruwash to the north and another at Zawyat Al Aryan to the south - correlate to Saiph and Bellatrix respectively, while three pyramids at Abusir farther south correspond to the head of Orion. Bauval and Gilbert also believe that the pyramids at Dashour, viz., the Red Pyramid and the Bent Pyramid, represent the Hyades stars of Aldebran and Epsilon Tauris respectively. Furthermore, this schema links Letopolis, located due west across the Nile from Heliopolis, with Sirius, the brightest star in the sky. The centre of this region, however, is Giza, known as the terrestrial equivalent of Rostau, the gateway to the Duat, or Underworld.

The region in Hopi cosmology corresponding to the Rostau is called *Tuuwanasavi* (literally, "centre of the Earth"), located at the three Hopi Mesas. Similar to the ground-sky dualism of the three primary structures at the Giza necropolis, these natural "pyramids" closely reflect the Belt stars of Orion. In addition, the portal to the nether realms is known in Hopi as the *Sipapuni*, located in the Grand Canyon. This culturally sacrosanct area mirrors the left hand of Orion. Whereas the Egyptian Rostau is coextensive with the *axis mundi* of the Belt stars formed by the triad of pyramids, the Hopi gateway to the Underworld in the Grand Canyon is adjacent to the centre place but still close enough to be archetypal resonant in that regard.

In a later book entitled *The Message of the Sphinx*, Robert Bauval and co-author Graham Hancock describe, among many other things, the cosmic journey of the Horus-King, or the son of the Sun, to the Underworld. "He is now at the Gateway to Rostau and about to enter the Fifth Division [Hour] of the *Duat* - the holy of Holies of the Osirian afterworld Kingdom. Moreover, he is presented with a choice of 'two ways' or 'roads' to reach Rostau: one, which is on 'land' and the other in 'water'*³*. We have been blessed with a wealth of hieroglyphic texts, both on stone and on papyrus, with which we can reconstruct the Egyptian cosmology. Unless we consider petroglyphs more as a form of linguistic communication than as rock "art," the Hopi and their ancestors, on the other hand, had no written language; hence we must rely on their recently transcribed oral tradition. In this regard the Oraibi *tawa-mongwi* ("sun watcher") Don Talayesva describes an intriguing parallel to Rostau. As a young man attending the Sherman School for Indians in Riverside, California during the early years of the twentieth century, he became deathly ill and, in true shamanistic fashion, made an inner journey to the spirit world. After a long ordeal with many bizarre, hallucinatory visions, he reached the top of a high mesa and paused to look. (Is it simply another coincidence that the Hopi word tu'at, also spelled tuu'awta, meaning "hallucination" or "mystical vision," sounds so close to the Egyptian Duat— indeed spelled by E.A. Wallis Budge, former director of antiquities at the British Museum, as *Tuat*, that seemingly illusory realm of the afterlife?)

> *"Before me were two trails passing westward through the gap of the mountains. On the right was the rough narrow path, with the cactus and the coiled snakes, and filled with miserable Two-Hearts making very slow and painful progress. On the left was the fine, smooth highway with no person in sight, since everyone had sped along so swiftly. I took it, passed many ruins and deserted houses, reached the mountain, entered a narrow valley, and crossed through the gap to the other side. Soon I came to a great canyon where my journey seemed to end; and I stood there on the rim wondering what to do. Peering deep into the canyon, I saw something shiny winding its way like a silver thread on the bottom; and I thought that it must be the Little Colorado River. On the walls across the canyon were the houses of our ancestors with smoke rising from the chimneys and people sitting out on the roofs."⁴*

In this narrative the narrow, dry road filled with cacti and rattlesnakes, where progress is measured by just one step per year, is contrasted with the broad, easy road quickly leading to the canyon of the Little Colorado River. A few miles east of the confluence of this river and the Colorado River is the actual location of the Hopi "Place of Emergence" from the past Third World to the present Fourth World. Physically, it is a large travertine dome in the Grand Canyon to which annual pilgrimages are made in order to gather ritualistic salt. In correlative terms the Milky Way is conceptualised as the "watery road" of the Colorado River at the bottom of the Grand Canyon - that sacred source to which spirits of the dead return in order to exist in a universe parallel to the pueblo world they once knew. This stellar highway is alternately seen as traversing the evergreen forests of the San Francisco Peaks, upon whose summit is a mythical kiva leading to the Underworld. Talayesva's account also includes traditional other-worldly motifs such as "the Judgement Seat" on Mount Beautiful, which supports a great red stairway, at least in his vision (this peak is actually located about eight miles west of Oraibi). Furthermore, we hear of a confrontation with the Lord of Death, in this case a threatening version of Masau'u (the Hopi equivalent of Osiris), who chases after him. Thus, like the Egyptian journey to the Duat, the Hopi journey to Maski (literally, "House of Death") has two roads - one on land and one on water. In this context we must decide whether or not the latter is really a code word for the sky. In the "double-speak" of the astral-terrestrial correlation theory, are these spirits in actuality ascending to the celestial river of the Milky Way? Is this, then, the purpose of the grand Orion schema? To draw a map on Earth which points the way to the stars?

Another startling piece of evidence from the Hopi oral tradition comes from one Jasper Poola of First Mesa. "Some Hopis have told me, for instance, that the original *Sipapuni* (Earth navel) was in Egypt, India, or Jerusalem—that is, somewhere in the Old World."[5] This suggests a considerably wider range of Hopi migration than hitherto acknowledged. Parallels between the Old World and the New can even be found in the field of comparative linguistics. For instance, the Hopi word *sohu* (or *soohu*) simply means "star," but in their belief system stars are conceptualised as supernatural entities, with those of Orion being ceremonially paramount.[6] The Hopi even have a Sohu *katsina* (also spelled *kachina*), an intermediary spirit whom dances wearing a mask with three stars in a row on its crest.[7] In the Egyptian Pyramid Texts, which are some of the world's oldest funerary literature, the word *Sahu* refers to "the star gods in the constellation Orion."[8] In addition, we find an important verification for the sky-ground dualism of the Orion Correlation Theory in both the homophone *Sahu*, which means "property," and its cognate *sah-t*, which refers to "landed property," "estate," "site of a temple," "homestead," or "environs." Because the term Sahu simultaneously refers to both stars and ground, this conceptual mirroring aligns the two realms, i.e., "...on Earth as it is in heaven." Other language correlations too numerous to explore here corroborate if not a Hopi migration from the Old World, then at least a pre-Columbian contact with Middle Eastern or North

African mariners, perhaps Phoenician or Libyan.[9]

Another linguistic synchronicity is found in the name *Arizona* itself. The first thing that comes to mind is the phrase "arid zone," which is a possible derivation for this term of uncertain origin. However, the Oxford English Dictionary cites a more germane meaning for the word "zone" as "The constellation named Zone or the girdle of Orion." In other words, a constellation within the larger constellation, otherwise known as his Belt. Following Aristotle's usage, the Latin poet Ovid refers to "Zona" specifically as the three central stars of Orion.[10] Considering the prefix Ari-, we readily think of Ariel, the powerful spirit of air and Earth in Shakespeare's play *The Tempest*. In the Old Testament Book of Isaiah, Ari-el is an appellation for the city of Jerusalem and literally means "Lion of God." In Ezekiel the word refers to the "Hearth of God," or the altar upon which burnt offerings were made.[11] Thus, in the name *Ari-zona* is hidden the fiery belt of Orion emblazoned upon one of the sacred navels of the world, in this case the three Hopi Mesas of the south-western United States.

Returning to the subject of Orion projected upon the deserts of Egypt and Arizona, we find both discrepancies and parallels. In terms of distinction, the Egyptian plan is on a much smaller scale than the one incorporating the Arizona stellar cities, using tens of kilometres rather than hundreds of miles. Furthermore, the bright stars of Betelgeuse and Rigel are perplexingly unaccounted for in the Egyptian schema. In addition, from head to foot the Giza terrestrial Orion is oriented Southeast to Northwest, while the Arizona Orion is oriented Southwest to Northeast. Of course, the pyramids are located west of the Nile River, while the Hopi Mesas are located east of the "Nile of Arizona,"[12] i.e., the Colorado River. We should also point out that Abusir is not in the correct location to match Orion's head on the constellation template. Bauval and Gilbert state that Abusir is "...a kilometre or so south-east of Zawyat Al Aryan."[13] (Corresponding to Bellatrix, or Orion's left shoulder); it is in fact about six kilometres Southeast. In other words, Abusir is nearly four miles south-Southeast of where it should be according to the Star Correlation Theory. Unlike Bauval, Gilbert, and Hancock, the present author has not yet travelled to Egypt, but the consultation of any scale map will verify this statement.[14]

Despite these few differences, the basic orientation of the Egyptian Orion is similar to that of the Arizona Orion, i.e., south, and the reverse of the celestial Orion. According to Dr. EC Krupp, Director of the Griffith Observatory, this is one of the many factors that invalidate the Orion Correlation Theory.[15] This critique, however, is the result of a specific cultural bias in which an observer is looking down upon a map with north at the top and south at the bottom. Imagine instead that the observer is standing on top of the Great Pyramid (or for that matter, at the southern tip of First Mesa) and gazing southward just after midnight on the winter solstice. The other two pyramids (or Mesas) would be stretching off to the Southwest in a pattern that reflects the Belt of Orion now achieving culmination in the southern sky. We can further imagine that if the upper portion of the terrestrial Orion were simply lifted perpendicular to the apparent plane of the Earth while his feet were still planted in the same position (Abu Ruwash and an undetermined site in the case of Egypt. Canyon de Chelly and Betatakin in the case of Arizona), then this positioning would perfectly mirror Orion as we see him in the sky.

When the Anasazi gazed into the heavens, they were not looking at an extension of the physical world, as we perceive it today but were instead witnessing a manifestation of the spirit world. Much like the Egyptian Duat, the Hopi Underworld encompasses the skies as well as the region beneath the surface of the Earth. This fact is validated by the dichotomous existence of ancestor spirits who live in the subterranean realm but periodically return to their earthly villages in the form of storm clouds bringing the blessing of rain. Even though the eastern and western domains ruled by Tawa (the Sun) remain constant, the polar directions of north and south, controlled by the Elder and Younger Warrior Twins (sons of the Sun) respectively, are reversed. Thus, the right hand holding the nodule club is in the east and the left hand holding the shield is in the west, similar to the star chart. However, the head is pointed roughly southward rather than northward. This inversion is completely consistent with Hopi cosmology because the terrestrial configuration is seen as a reversal of the spirit world, of which the sky is merely another dimension.

At any rate, looking up at Orion on a midwinter night, we can imagine that *our* perspectives have switched and that we are suspended high above the land. Suspended gazing from the Northeast to the Southwest toward the sacred *katsina* peaks and the head of the celestial Masau'u suffused in the evergreen forests of the Milky Way. Ironically, it is here on the high desert of Arizona that we also intuit the truth of the hermetic maxim attributed to the Egyptian god Thoth (Hermes Trismegistus): "As above, so below."

© 2000 - Gary A. David - http://azorion.tripod.com

FOOTNOTES

1. *Grigsby cited in Graham Hancock and Santha Faiia's book Heaven's Mirror: Quest For the Lost Civilisation (New York: Crown Publishers, Inc., 1998), p. 127.*

226

2. Robert Bauval and Adrian Gilbert, *The Orion Mystery: Unlocking the Secrets of the Pyramids* (New York: Crown Publishers, Inc., 1994), p. 125 ff.

3. Graham Hancock and Robert Bauval, *The Message of the Sphinx: A Quest for the Hidden Legacy of Mankind* (New York: Three Rivers Press, 1996), p. 175.

4. Don Talayesva, Leo W. Simmons, editor, *Sun Chief: An Autobiography of a Hopi Indian* (New Haven: Yale University Press, 1974, 1942) pp. 121-128.

5. John D. Loftin, *Religion and Hopi Life in the Twentieth Century* (Bloomington: Indiana University Press, 1994, 1991), p. 69.

6. Edward S. Curtis, Frederick Webb Hodge, editor, *The North American Indian: Hopi, Vol. 12* (Norwood, Mass.: The Plimpton Press, 1922), p.101.

7. Barton Wright, paintings by Cliff Bahnimptewa, *Kachinas: a Hopi Artist's Documentary* (Flagstaff, Arizona: Northland Publishing, 1990, 1973), p. 125.

8. E. A. Wallis Budge, *An Egyptian Hieroglyphic Dictionary, Vol. II* (New York: Dover Publications, Inc., 1978, 1920), and page 638.

9. Barry Fell, *America BC* (New York: Demeter Press Book, Quadrangle/The New York Times Book Co., 1977, 1976), pp. 174-177

10. Richard Hinckley Allen, *Star Names: Their Lore and Meaning* (New York: Dover Publications, Inc., 1963, reprint 1899), p. 315.

11. William Smith, LL.D., *Smith's Bible Dictionary* (New York: Family Library, 1973), p. 49.

12. Jefferson Reid and Stephanie Whittlesey, *The Archaeology of Ancient Arizona* (Tucson, Arizona: The University of Arizona Press, 1997), p. 112.

13. Bauval and Gilbert, *The Orion Mystery*, p. 139.

14. T.G.H. James, *Ancient Egypt: The Land and Its Legacy* (Austin: University of Texas Press, 1989, 1988), p. 41.

15. EC Krupp, *Skywatchers, Shamans, & Kings: Astronomy and the Archaeology of Power* (New York: John Wiley & Sons, Inc.), pp. 290-291.

Now, you may see that this knowledge was universal amongst the ancients of our world on both sides of the Atlantic Ocean. Orion was the great navigator, a moment in time and God on Earth, in ancient days. His position in the sky heralded Taurus the great Bull, the constellation that was required for transatlantic passages by the navigators of those distant times. The evidence of Gary David clearly shows that there was the same knowledge on both sides of the Atlantic. So let us now move forward to the methods used by the ancients to cross the Atlantic Ocean.

The rules for following a Star across the Atlantic

1. Follow a star that is over 17° latitude at 17° altitude and head South West following the course of that star using triangulation methods .

2. Keep watch with an hourglass using Las Palmas time that was acquired from the local Watcher before departure.

3. Keep the ship's watch with a separate hourglass on observed local time against Las Palmas time.

4. Take 22.00 sightings on the Pleiades.

5. Take midnight sightings on Orion.

6. Take noon sun sightings, using the laws of equal altitude to find true south.

7. Take lunar sightings if available.

8. Subtract Las Palmas time from the ship's local time and calculate position of longitude.

9. Check speed through the water using dead reckoning as a support to astro navigation.

10. Estimate speed of current.

11. Mark Estimated Position (EP) on vellum chart and update daily.

12. Maintain Log.

13. Continuously monitor the ship's latitude against 17° latitude.

14. On reaching the latitude of 17° turn westward and follow that line of latitude without deviation.

15. Consult an almanac for stars that should be due south of your position at other times of night.

16. Arrive successfully at Nelson's Dockyard in Antigua at 17° latitude.

The return journey would not be made until the spring and would then take a route around the north side of the Sargasso Sea, via the Gulf Stream and The North Atlantic Drift. They would arrive on the North European coastline anywhere from the Outer Hebrides to Lisbon depending on the starting latitude and the star they followed.*lxxxviii*. If a navigator departed from the Caribbean and sailed North on the Gulf Stream along the eastern seaboard of America until he reached the latitude of 46°N, the ship would end up located at Scatarie Island off the coast of Cape Breton in Newfoundland. This area is reputed to have been visited by the Vikings who tried to establish a settlement around 1000 AD but failed. St Brendan the Abbott is believed to have set sail from Ireland in a carrach in the hope of reaching the far side of the Atlantic, and actually landing in Newfoundland. Also, there is evidence that the Portuguese Cod Fishermen made regular crossings of the Atlantic in pre-Columbian times. By following the 46° latitude line east, the ship would avoid the Sargasso Sea and arrive at La Rochelle in the Bay of Biscay on the Coast of France.*lxxxix*

La Rochelle is the area that the Knights Templar fleet escaped from after the capture and torture of their leader, Jacques de Molay. We will cover the evidence of the Knights Templar's navigational abilities further on. Is all this evidence a co-incidence, as the academic institutions would have us believe? Or does it reveal a hidden knowledge inherited from a distant past and prove that mariners from a pre-Columbian age not only knew of these lands but more importantly, knew how to reach them? Is there further evidence that this sea route was used in prehistoric times? The following discovery of what may be the oldest sea charts in the world may well lead you to consider that there is such evidence.

THE OLDEST SEA CHARTS IN THE WORLD?

Andrew Collins in his latest book '*Gateway to Atlantis*', included a section on Bovine Art in ancient Cuba and showed two pictographs from Aboriginal Archaeology in Cuba by Ramon Dacal Moure and Manuel Rivero de la Calle, published in 1986.[xc] During this narrative, he questions the meaning of the cave drawings, since there is no prehistoric record of bulls on the island. One of the pictographs cannot be assigned a location, but the other is from the Ceuva del Aguacate at Guara in the province of Havana. Andrew Collins speculates that they may depict hunting scenes, bullfights or bull baiting, but I believe it was demonstrating something entirely different, something of much more significance. It is my belief that they are the simplest and oldest ancient sea charts ever found, and only a navigator could understand them. In my mind, they not only show the direction in which to voyage safely, east or west, but also the season to embark on such a journey.

The bull has been depicted in an east facing direction. Taurus the Bull, the constellation sign on the ecliptic also faces east when translated onto the Earth. In the winter, due to the angle of the ecliptic, the Bull constellation does appear to be angled upward, just as it has been shown in the Cuban cave drawing. The declination of the constellation at that time of the year would cover the Sargasso Sea when over the mid Atlantic. It should also be noted that the figure on the right is in the same position on the zodiac as Orion.

The Sargasso Sea at the latitude of 21° to 44° and longitude 41° to at least 74° has to be avoided at all costs by sail driven ships. It was a terrible place to die a slow lingering death, with no winds, thick seaweed, flotsam and jetsam and at 2000 miles across, anyone caught in this section of the sea had no chance of escape or survival. The sea was depicted in earlier times as a spiral, designed to draw the unwary into its central core as the inner edges of the ocean currents caused eddies. Strangely, this spiral is also depicted on the north stone of the ancient Neolithic circle at Temple Wood at Kilmartin in the West of Scotland, with one of the spirals facing west. Later in Mediaeval sea charts, this region of the sea was illustrated with sea monsters in the hope of scaring away the careless navigator. More likely however, is that the intention was to frighten off the trading competition. What is remarkable is that it is approximately 30° of longitude, which is the same as Taurus was in the ancient autumn sky.

The figure on the left of the pictograph appears to have a black face and the one on the right a white one, which I believe is depicting people's colouring at lower and higher latitudes. The figure on the left is indicating a direction of travel northeast over Taurus and the figure on the right is indicating southwest. Both of these figures are indicating that the navigator must clear Taurus on the sea, north and south. Following this advice would steer a navigator away from the dreaded Sargasso Sea as well as putting his ship firmly on the trade winds and circulating oceanic currents. To follow these routes would ensure a landfall either side of the Atlantic. To carry out such an exercise in navigation, a method such as the cross would have to be employed. The drawings that are shown are a computerised impression of the pictographs and are as nearly accurate as I can make them.

There is a further pictograph that also appears to be illustrating the star sign of Taurus, again facing east. In front of Taurus is another white faced stick man who seems to be running west and between the white-faced stickman and Taurus are four diagonal lines. I believe that these four diagonal lines are extremely important and that they are depicting 30° of latitude. To the right of these four diagonal lines are two lines like a path, running north east to south west, with what appears to be an arm along the path, with a pointing finger directing the runner south west. We already have knowledge of that fact that the ancient navigators travelled by the stars, so this pictograph makes perfect sense as a representation of a man on the European side of the Atlantic being directed which route to take to the Caribbean. So why is the depiction illustrating the action of running? Does this imply flight or escape? At 30° north of the Cape Verde Islands, is the place and seaport named La Rochelle in France. The distance on the meridian in degrees of the constellation of Taurus is 30°.

La Rochelle was the place where the Templar fleet fled from the arrests instigated by the then Pope, Clement V on Friday 13th October 1307. It is written that the whole fleet slipped away never to be seen again.[xci] Many of these fleeing Knights settled in Scotland and others settled in Portugal. Both of these countries were free of the rule and constraints of the Roman Catholic Church. The month of their departure is also significant, October is of course, the right time of the year to be making a voyage across the Atlantic to the Caribbean from Europe. Could this pictograph be a depiction of that escape?

It cannot be ruled out that this is a diagram illustrating the escape of some of the Knights Templar. There were many candidates for potential escapes from tyranny who may have had the knowledge that I have discovered. According to Andrew Collins even the Portuguese had a legend of seven men and their families escaping by travelling westward across the ocean.

The Essenes may well have been another group of gnostics who escaped from the Roman's destruction of Qumran in this fashion, leaving the Dead Sea Scrolls behind them. The Essenes translates into the School of Fishes, after the new age of Pisces. There are statues and memories amongst the indigenous Amerindians of a group who they called the fish people, and some stone carvings depict half men half fish. Were these the fleeing Essenes?

Andrew Collins goes on to say that in the Cueava de Ambrosio, near Varadero there is "a Christian Priest stickman" drawing located next to a Calvary cross.[xcii] He debates whether or not it is in the style of Taino or African artists. I believe that it is in fact older than the age of these artists and is actually further evidence of the cross being used as an instrument for navigational purposes in pre history. Throughout history, there is much evidence that this mathematical knowledge and the use of the cross is cyclical and resurrects on a regular basis.

Certainly the recent time line of cyclical rediscovery seems to be at around 1000-year intervals at around the turn of the millennium and at a time when spirituality seems to be declining. Christ changed everything including the Roman Empire around the year 0 and after some time the knowledge appeared to disappear again in Europe and was not resurrected again until after the Crusades in the year 1095BC. It has disappeared again until now. The people who resurrected the knowledge again in the last millennium were the Knights Templar and left us a rich but highly secret set of clues.

CHAPTER 8

THE KNIGHTS TEMPLAR

The Templars were a poor chivalric order who disdained materialism until their arrival in Jerusalem in the early twelfth century on a papal crusade. Eleven knights dug under the ruins of Solomon's Temple for nine years and hit pay dirt. Under the Temple Mount the Templars discovered a network of tunnels where the earlier Jewish priests had hidden their treasures from the marauding Romans in 70 AD. The knights had not only uncovered some of the gold and silver described in the Copper Scroll of the Dead Sea Scrolls, but also the records of the first Jerusalem Church of Christianity. This, according to Knight and Lomas in their book '*The Hiram Key*', is precisely what they went digging for.

The first Bishop of the Jerusalem Church was Jesus' brother James. He was assisted by Jesus' other siblings of five brothers and two sisters. After the Romans destroyed Jerusalem and dispersed the Jews, James and his family fled to France, a fact confirmed by the best-selling book '*The Holy Blood and The Holy Grail*' by Baigent, Lincoln and Leigh. These descendants of the holy Davidic line had the wherewithal to gain wealth throughout Europe and buy vast properties. Over time they constituted a large proportion of the European royalty.

Whilst professing whatever religion the state was demanding at the time, the Knights first worshipped the pantheon of Roman Gods and later papal Christianity, but all awhile, they secretly practised the Christianity of the early Jerusalem Church, the secrets of which they had excavated. These descendants of the priestly line, now went by the name of Rex Deus, and passed on their traditions throughout the royal courts of Europe where they ruled. The Rex Deus founded the original Celtic church, and St. Patrick of Ireland was a Jew. Other Rex Deus included the Habsburgs, the Stuarts of Scotland and today's King Juan Carlos of Spain, who includes among his titles, Protector of the Holy Places of Jerusalem.

The Templars returned from their Crusade with vast wealth uncovered from underneath the Temple Mount and embarked upon a monumental building program in Europe in the Middle Ages, which ultimately resulted in the construction of over 250 castles and Gothic churches. They were now in a position to challenge the papacy for political control of Europe, yet the usually jealous Vatican permitted their power to grow unabated. Knight and Lomas claim that this is because the Templars had an advantage over Rome.

According to Knight and Lomas' version of events, the Knights had uncovered the original records of the Jerusalem Church and they found that these records proved that the Vatican was not practising Christianity as its founders had originally intended. For example, a pope with exclusive rights of divine interpretation was not part of God's plan. If the scrolls were made public, they would jeopardise Rome's legitimacy.

In the fourteenth century, the French king known as Philip the Fair co-ordinated a purge of the Templars in conjunction with the Holy See. Their properties were seized and the leader, in fact the last Templar leader, Jacques Burgundius de Molay, was brutally tortured by crucifixion until he confessed to the alleged sins of his Order. He survived the first round of torture and a linen cloth was placed over him to absorb his blood and keep him warm. This cloth became known as the Shroud of Turin and the author's arguments for this claim are superbly grounded in fact.

A group of Templars led by the St. Claire clan escaped the French siege by ship and landed in Scotland, a safe port since it was ruled by the Rex Deus member, Robert the Bruce. It must be noted that at his death, Robert the Bruce's heart was transported to Jerusalem for burial. Two versions of what happened to his heart are in existence. The first claims that the deliverers of the Bruce's heart were captured in Spain, the second insists they made it and the heart lays buried in Jerusalem, still to this day.

According to Knight and Lomas, the French Templars brought with them the damning scrolls that they had discovered under the Jerusalem Temple Mount and built a chapel in Roslyn, near Edinburgh, specifically to house them. The chapel built by William St Clair (1415- 1482), that still exists in its entirety today, was built according to the specifications of the Herod Temple and an exact duplicate of the Western Wall was constructed alongside it. To hide their now-banned Templar roots, the Knights formed a new organisation known as Freemasonry, whose rites were precisely those as laid down in the controversial scrolls.

Scottish rite Masonry rapidly spread throughout Europe and according to Knight and Lomas, it was now much to the chagrin of the Vatican. In Knight and Lomas' version of events, the pro-Catholic Hanoverian dynasty of 1717 took over the British crown and ordered vast changes in Masonic rites. The 33 Scottish degrees were reduced to just four in the new English rites and all the secrets that were incorporated within the Jerusalem scrolls were eliminated altogether from the lodge ceremonies.

However, Knight and Lomas' views are not agreed with on the whole, other writers share a different version of events, insisting the Templars and the Vatican remained a partner throughout their public trials. F. Tupper Saussy in his excellent book, '*Rulers of Evil*' writes:

> "*On March 22, 1312, Clement V dissolved the Knights Templar with his decree Vox clamantis. But the dissolution proved a mere formality to further appease Philip (the Fair). More importantly, it permitted the Templars, in other manifestations to continue enriching the papacy...A subtle provision in Vox clamantis transferred most Templar estates to the Knights of St. John of Jerusalem. In Germany and Austria, the Templars became "Rosicrucians" and "Teutonic Knights." Six centuries later, as the "Teutonic Order," the Knights would provide the nucleus of Adolph Hitler's political support in Munich and Vienna.*"

On January 26, 2001, the Israeli Government's television outlet, Channel One, broadcast a video clip secretly filmed under the Shrine of Omar, the 7th century Islamic building which may have been deliberately constructed over the Holy of Holies, the most sacred prayer room of the ancient Temple. The video revealed a new and massive tunnel aimed directly at the most sacred core of Solomon's and later, Herod's Temples. On Jan. 4, 2001, Colin Grant published a full front-page article in the Glasgow Sunday Post. He revealed that the Knights Templar, as descendants of the Davidic line, believe they have the right to reclaim the Temple Mount as their true heritage. Now you may begin to see why some believe there is a bloodline still in existence today, and that this Davidic bloodline goes right back to Egypt through the age of the Merovingian Kings of Burgundy and Mary Magdalene to the ancient House of Amen. Well, it is hard to logically deny that this is a distinct possibility even in the most stubborn of minds, especially in the light of the latest scientific revelations of genetics and the knowledge that bloodlines do exist. The descendants of the House of Amen either exist on the Earth here and now, or the bloodline died out somewhere in the past. Much of the bloodline may now be mixed and other genetic traits influence the individual offspring, but it is surely there to be found.

What may be harder for the reader to understand, is that these people don't just believe in the reincarnation of the material bloodline and genetic make up of inherited traits. They also believe in the reincarnation of the soul, character and mind of the people who lived long ago. The stage of the latest actors is different because of the Golden Thread of Time, but the characters are the same. There may be those who lead us and govern our lives who believe this and run affairs in line with these beliefs and prophesies, and whose sole intent is to set up a world government based on ancient ideals and prophesies. The proof that this powerful belief system existed in the Middle Ages through the Knights Templar, over a thousand years after the crucifixion of Christ, is to be found at Roslyn Chapel, near Edinburgh in Scotland.

ROSLYN CHAPEL

Located at N55° 49' 026" by W003° 05'653" at an altitude of 613 feet above sea level Roslyn is considered to be on the rose line near the Pentland Hills which lie to the west. Roslyn is a word that comes from a maritime heritage and means rose line, which is a name connected to the compass rose as shown on modern and ancient charts. The Pentland hills are named after the pentagram, which is the five-sided shape sometimes associated with witchcraft, but as we now know, is a mathematical system of measurement and triangulation. To the south of Roslyn, lies the original village of Roswell.

The western entrance to Roslyn Chapel with a rose window and castellated cross over the door.

This village, castle and chapel has long been associated with ancient, sacred knowledge and the area housed a library and printing press in ancient times that produced rule books for the guidance of Princes throughout Europe. The knowledge that transcends the normal is still in evidence today, as the laboratory that created Dolly the Sheep, the world's first cloned animal, is located within the bounds of the village. Roslyn Chapel is an ancient chapel built by the St Claires and Knights Templar and it was designed to be a representation of The Holy City of Jerusalem. Some consider that the hymn Jerusalem is a reference to Roslyn Chapel. Roslyn Chapel is a collegiate church, which means that it is acoustically designed to enhance the performance of choirs who used to perform music there. King Edward VII visited and stayed at the Inn next door, which had many other distinguished visitors including Dr Samuel Johnson, Alexander Maysmith, Sir Walter Scott and William and Dorothy Wordsworth. Her Majesty the Queen of England and the Duke of Edinburgh visited the chapel on the 29th of June 1961. Prince Charles, the Prince of Wales visited in 1998.

It was built in 1446 AD, a number of years before the discovery of America by Christopher Columbus. It was decorated with carved American Maize plants and Aloe, and it is a truly remarkable representation of the stonemason's craft. Many masons visit it today and see the symbols of their society recorded in stone workings of such ingenuity as to stagger the modern mind.

American Maize in Roslyn Chapel

Below the main structure at the eastern end, is the so-called "Leper Chapel", probably named as such to keep the curious away. The nearest leper colony was at Liberton or 'lepertown', some miles north of Roslyn. In this chapel there is a small antechamber, where there can be found the marks made by the Mark Masons who constructed the building under the direction of Knights Templar. To the left of the door lintel can be seen quite clearly a line carving of a cross on a tripod amongst other marks that look like stars. Stars are everywhere within the chapel, especially upon the ceilings. There is also an illustration of the St Claire Cross, which is castellated and is carved into the ceiling as well as being represented in wood. The stained glass window facing east represents the figure of Christ with a blue pyramid as a halo.

THE TEMPLAR TREASURE WAS NOT FOOL'S GOLD

All the ancient pagan and biblical symbols are represented in some form within this remarkable building. It is also home to the best museum of masonic regalia and pictures that are available today. In some of the pictures that are on display representations of the quadrant and square can be seen. There is also a whole area dedicated to the Gypsy or "Egyptian" people as well as the Black Madonna confirming connections to Egypt and Mary Magdalene who was reputed to have escaped to France after the crucifixion of Christ.

There is evidence in the carvings that the Knights Templar were trying to express the same knowledge of *"The Golden Thread of Time"* that the ancient Egyptians and the House of Amen had tried to impart thousands of years earlier. There is a carving by the altar, which shows the fallen angel that has become bound to the Earth and materialism, and the meaning behind this idea is that it is Satan or Set who binds us in earthly form within ourselves. The war between chaos and order is in our own cosmic minds.

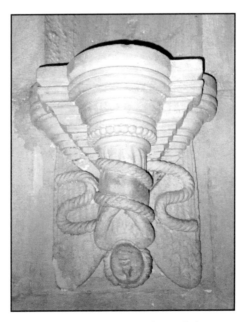

*The fallen angel is shown to be upside down
and above the vine of life.*

THE VINE OF LIFE

The *"Vine of Life"* is carved into friezes in many mediaeval cathedrals and churches as well as in Roslyn Chapel. The importance of this image is to impart to the initiate, the understanding of their place and completeness within the *"Golden Thread of Time"*. The craftsmen were trying to explain the true nature of time with the vine of life. They were attempting to explain that no one is separate from the One and that all individuals are connected to Nature, each other, and the endlessly repeating life force. This is understood more simply by the explanation that no human being is created separately, but the genes and instructions that make up the body of the individual living person are passed down through mother after mother from the beginning of time to the present moment. If an individual does not reproduce, it does not matter, as the vine also spreads many branches, which will bear fruit. This is where the teaching that all men and women are brothers and sisters comes from and it is time related. This illumination is shown to the initiate on the Apprentice Pillar at Roslyn Chapel.

On the Apprentice Pillar at Roslyn Chapel, The Vine of Life is seen issuing from the mouth of the serpent.

The pillar is shown with serpents and the vine of life intertwined in the same manner as depicted on Celtic crosses. The concept is to show the spinning cyclical nature of time, the interconnected nature of all things and the binding of spirit to the Earth by the serpent or dragon which is representative of the ruthless spirit of Nature.

ENGLAND'S ST GEORGE AND THE DRAGON

In the county of Lincolnshire, England, is the great building of Lincoln Cathedral, which is reputed to be one of the finest constructions of its type in Europe. Lincoln Cathedral was built in 1072 by the Normans and consecrated in 1092. After a destructive earthquake in 1185 it was rebuilt; and Fleming Chantry added the spires in 1420. High on a hill and overlooking the country-side below, this east aligned and cross-shaped masterpiece of medieval archi-tecture embodies many similar concepts to the carvings and stories of Roslyn chapel. This great building and seat of the Holy See is also designed as a time piece that tries to relate cosmic ideas to the initiate in the same way. The French knights from Normandy, conquered England in the famous battle of 1066 at Hastings. The location of Hastings is just along the coast from the Long Man of Wilmington discussed earlier in the book. The Norman knights were also responsible for the 1092 massacre of the Moslems at the infamous siege of Jerusalem. It was the burning of the Holy Church of the Sepulchre in Jerusalem by an Egyptian King that started the Crusades, which were led by these French knights. It was the French Knights Templar who excavated the Temple Mount before Saladin re-took Jerusalem. It is the friezes of St Hughes Choir near the Norman Apse, in the very heart of Lincoln Cathedral, that carries the secret of the Vine of Life and the Serpent.

The frieze of St George slaying the dragon in Lincoln Cathedral.

In the frieze of St George and the Dragon, one can see the message that was being attempted by the stone masons to describe the slaying of the serpent in the Vine of Life by St George. This is also represented as being a cyclical event as there are many images of St George slaying the dragon at regular intervals along the Vine. Like the Hindu Baghavit Gita, this shows death, rebirth and the cyclical battle against the material forces that govern our earthly bodies. That battle is fought, won or lost on an individual basis, but also within nations and civilisations. All civilisations grow with enthusiasm and fervour and then become corrupted and die. Our western civilisation is approaching that position now. It is only mathematics and cycles after all and quite predictable.

THE GREEN MAN

Many of these medieval buildings show representations of the Green Man and Roslyn is no exception to the rule. The Green Man, which shows the Vine of Life growing through the head and mouth of an individual, is considered to be a pagan symbol and associated with fertility. It is thought to be pre Christian, but it is not so because Christ himself is said to have stated the he was the "*Vine of Life*". The Green Man, thought by many to represent a demon, is often shown to be distressed, confused and angry.

This is an idea that is related to the Hindu concept of the Wheel of Life and the endless re-absorption of the material body back into nature as the vine is seen to issue from his mouth. It should also be remembered that they saw that new life could only be made possible by the death and decomposition of old life. The teaching that is being related to the initiate is again connected to the "*Golden Thread of Time*" and the Vine of Life.

No wonder the Green man looks distressed, confused and angry. He is trapped on the constant cycle of death and rebirth and doesn't understand what it is all about. A good example of the Green Man may be seen in Roslyn Chapel as well as Southwell Minster near Nottingham.

There are writings on the graves outside Roslyn Chapel, which clearly reveal the attitude they had toward life and death even recently. These are the different virtues and the beauty of character to be seen on the mausoleum of a Lady.

"*Patience, Faith, Hope, Love, Courage and Truth.*"

Here is a beautiful writing containing the nature of time and the endurance of Love throughout.

> *"Not stone or brass, these perish*
> *With the flight of time and quickly pass.*
> *But love endures in every time.*
> *Eternal as the poet's rhyme".*

Another one that is to be seen almost worn away on a grave on the western side of Roslyn Chapel states the concept of death in the eyes of the enlightened.

> *"The souls of the just are in the hands of God and the torment of*
> *death shall not touch them. In the sight of the unwise, they seemed to*
> *die, but they are in peace."*

They had no fear of death because of their personal understanding of nature and its laws, rather than relying on blind faith. It is the cross that brought this spiritual enlightenment to the Knights Templar because they became aware of the nature of time, and as you will see from this second poem, there was a nautical connection to the words that were written.

> *"Safe, safe at last from doubt, from storm, from strife,*
> *Moored in the depths of Christ's unfathomed grave.*
> *With Spirits of Just, with dear ones lost and found again.*
> *This strange ineffable life*
> *Is life eternal.*
> *Death has here no place. And they are welcomed best who suffered*
> *most".*

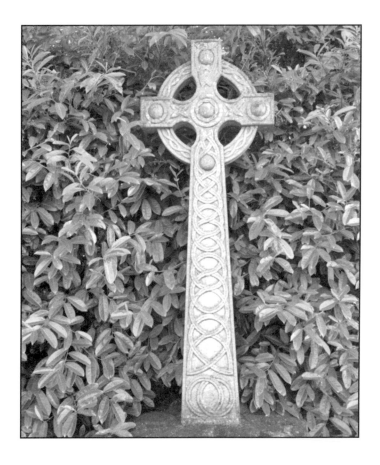

THE VINE OF LIFE, THE SERPENT AND THE CROSS

The image of the Celtic cross above is taken from the church of St Mary the Virgin at Great Bennington, Northamptonshire, England. This is the church of the Spencer family and is the last resting place of some the relatives of Princess Diana, including her father. On this cross you will see the vine of life intertwined and the serpent around the wheel, but you will also see that there is a hub at the centre as well as markers at the end of each cross piece. For not only is this cross a spiritual instrument, but it is also a practical one. It is not enough to have just one side of the instrument functioning without understanding the whole of its full capability. Like our brains, which are both practical and creative, to be a complete and balanced person both sides have to work in harmony. So it is with the cross, as it is both practical and spiritual in function. Using reason and common sense, the ancient peoples worked at achieving this balance and passed the knowledge down to us so that we might also benefit from their logic. The nature of time and place in the material world and the world of spirituality was originally conceived through navigation and astronomy. The Knights Templar recovered this knowledge, firstly on the practical side, and travelled the oceans of the world in secret. That they also navigated the spiritual world is clearly shown in their many works all over the world.

There is a legend in Scotland that claims Henry St Claire had sailed to America in pre-Columbian times and landed in what is now known as Boston, Massachusetts. How did he manage to navigate his way across the ocean? I believe that William St Claire may have had the cross and plumbline inherited from the knowledge of the Knights Templar when they excavated the Temple of Solomon several hundred years earlier. They also had the benefit of sea charts that they had discovered at the same time, which superimposed the stars onto the Earth's surface creating a system of gridlines or Ley Lines that located fixed places on the Earth with great accuracy. It is my belief that they resurrected the old and abandoned sites and re-surveyed the world to coincide with the new positions of the stars in relation to the changes caused by precession over the millennia. However, this system could only be made to work with an instrument like the cross. Suffice to say that I believe the real golden "*treasure*" that they found in the seven years of excavating Solomon's Temple, enabled the Pauper Knights to develop into a wealthy and influential power.

Their finds were not treasure in the conventional sense, i.e. it was not the gold and jewellery that some believe it to be, but was in fact, the discovery of the cross and knowledge of the American Continent. Plus the all important information on how to reach this far flung foreign land and trade with the descendants of the ancient mariners who had settled there thousands of years before.

The Author observing the glory of the rising of the sun on the ecliptic path, at the summer solstice 2001, using a hand held working cross.

Observe that the cross is slightly off the horizontal so that the wheel which is weighted and on a hub is to be seen as off centre.

The angle may be read directly from the scale that is around the wheel, through the hole at the bottom of the upright.

The picture above of a small hand held working cross is marked out in degrees with sight lines and viewing holes at each cross bar so that the viewer can read the angle. Unlike the stone one on the grave earlier, this cross does not celebrate death but life, as the Earth is renewed on a new day at midsummer when the Aten (Ra) as the Egyptians called it, is rising at its highest point on the eastern horizon. I took this photograph on the Scottish borders on my way to Roslyn Chapel. There was a chill in the still air and dew on the grass, but already the birds were starting their dawn chorus and the first crows were leaving the pine trees behind me to forage for their morning meals as they had always done on this day for countless generations. All of the wild life around me was stirring and celebrating the end of darkness as the first fingers of the sun's rays lit the underneath of the clouds along the eastern curve of the Earth and the promise of a renewal of life was fulfilled. The base of the wheel of this cross is a pendulum and freewheels on a hub, so that the heaviest bottom part of the wheel, which is marked at 0°, is always pointing to the centre of the Earth. Each side of the cross is marked to 90° allowing observation in each direction along the horizon.

By observing the positions of the circumpolar stars before dawn, it is possible to predict the point and time on the eastern horizon at which the sun will rise. As you will notice, there are two distinctive hills in the distance, and by predicting the correct spot to observe the sun rise on the summer solstice between these hills, would allow the instant establishment of a permanent ancient observatory. What we see in following the cyclical history of the cross and its nautical associations, is a constant reincarnation of the ancient knowledge throughout the ages. The cross, in its many forms, was in the hands of the Palaeolithic people before the ice age ended. The survivors of the flood passed it down to the Neolithic people and from there it was in the hands of Phoenicians, Minoan, Celts, the people of India, the Egyptians, the Hebrews, the Romans, the Essenes, the Inca, Tolmec, Maya, Hopi, Aztec and the Knights Templar.

Its latest reincarnation was in 1998 when it was born again in my mind and I predict that its recognition will become public worldwide knowledge for the first time ever, in a generation or two. I believe that the last time this knowledge disappeared from the most of the world was during the destruction of Alexandria.

ALEXANDRIA

This information had probably previously been deliberately destroyed in Alexandria with the burning of the Library in 500 AD by the Christians. Alexandria was the legendary library founded by Alexander the Great and built by his Greek general, Ptolemy I, King of Egypt, and his son, Ptolemy II, Shelley's Ozymandias. A showcase of classical learning, the Alexandrian library was originally housed in the Mouseion, an arts and science complex that included laboratories and conservatories, as well as a zoo and the library itself. At its height during the third century BC, the library was said to have housed over 700,000 papyrus manuscripts. Its librarians, among them Archimedes and the astronomer Aristarchus, had collected the works of, among very many others, Plato, Aristotle, Thucydides, Sophocles, Euripides, Hippocrates and Euclid, often when these authors were still at work.

As well as storing books, the library also acted as a publishing house, copying manuscripts for distribution throughout the ancient world. It aimed to house every book ever written, and thus being home to "all the knowledge in the (known) world". Originally founded in 332 BC, Alexandria developed into one of the greatest and most famous cities of the classical world and continued to be the capital of Egypt until AD 969, although its decline had set in several hundred years before. One thousand three hundred years of culture and study ensued.

There is a quote by the Greek scholar Athenaeus that sums up what the Alexandrian library was all about.

'And concerning the number of books, the establishment of libraries,
and the collection in the Hall of the Muses, why need I even speak,
since they are all in men's memories'

Athenaeus was not wrong. The library of Alexandria has inhabited the memory of man for close to two thousand years now. So, what was all the fuss about?

The Roman chronicler, Livy, claimed that 40,000 volumes were torched and destroyed when Julius Caesar decided to burn the fleet of Cleopatra VII in Alexandria harbour. The volumes were supposedly housed in grain depots near the harbour. But let's back up a bit and search for the origins of this amazing place. The first mention of the library is in 'The Letter of Aristeas' (circa 180 - 145 BC), a Jewish scholar who was housed at the library and was chronicling the translation of the biblical Septuagint into Greek, by the seventy-two Rabbi's attendant. The Athenian (in exile) Demitrius of Phaleron had commissioned this huge undertaking under his then patron, Ptolemy I, Ptolemy Soter. Demitrius had himself been a scholar and a bit of a tyrant in Athens, hence the exile, and had been a student of Aristotle himself. He had known Alexander the Great and was a first generation Peripatetic scholar of the highest order.

Another of Aristotle's students was Theophrastus, who with the backing of Demitrius, established a lyceum devoted to the studies and writings of Aristotle and modelled after Plato's academy. When in 297 BC, Theophrastus was invited to tutor Ptolemy's heir (Ptolemy having established himself as Pharaoh of Egypt), he declined and instead recommended Demitrius, who had recently been driven out of Athens in a political situation that was uncertain due to the wrangling of Alexander's successors.

It seems that Demitrius recommended that Ptolemy should gather together a collection of books on kingship and ruling, in the style of Plato's philosopher kings.

This he did, but he also gathered together books from all the then known world peoples, according to the commentator Aristeas, so that he could better understand subjects and trade partners. It would seem that Demitrius was also influential in the founding of a museum in Alexandria and of a temple to the Muses, the Greek divine patrons of the arts and sciences. Alexandria was the most prosperous place in the Mediterranean at this time, it was a rich centre for trade and commerce and ripe for a cultural flowering. At the crossroads between east and west, it attracted the very best thinkers of its day and an amazing gathering of minds constantly thronged its streets. Scholars were invited to come and carry out such peripatetic activities as observation of the heavens, mathematical deduction, medicine, astrology, geometry and history. It remains a well known fact that for the next five hundred years; most of the western world's discoveries were made there.

No remains of the museum which housed the original library have been found by archaeologists yet, but portions of the so called 'daughter library' have been unearthed in the temple of Serapis, in the northeast sector of the city. This was probably in the palace grounds and was surrounded by courts, elaborate gardens and a zoological park containing exotic and rare animals from around the Alexandrian Empire of the time. The chronicler Strabo said that in its centre was a great court and large circular dining hall with a dome. This domed hall contained an observatory in its upper terrace and had classrooms and outlying halls surrounding it. The style and layout is reminiscent to the Serapeum, which was built by Ptolemy II, Philadelphus and completed by his son. Some 50 scholars were housed there, with the royal family as their patrons, though later they were to funded by public money. In all probability, the actual shelves of the library were in one of the outlying halls or in the garden. These mainly consisted of pigeonholes in the racks for scrolls and parchments, many of which were wrapped in leather and linen. Later in Roman times, manuscripts were written in codex, or book, form and stored in chests known as armaria. It was a hundred years after the inception of the library that Aristeas wrote his 'letter'. In it he records that Ptolemy I handed over the job of gathering scrolls and books to Demitrius. He also attended to supervising the ongoing effort to translate many of the foreign works into the Greek language. This is why the seventy-two Rabbis' were translating the biblical Septuagint of the Old Testament mentioned earlier.

Greek libraries in this age were usually personal collections of manuscripts and scrolls, a good example of this would be the personal library of Aristotle, which housed his own as well as others' work. The temples of Egypt no doubt must have housed a treasure trove of manuscripts and documents and it seems to have been the desire of Ptolemy I to gather as much of this as he could. It was this desire and the efforts of his translators that made the library a truly huge undertaking. Ptolemy I wanted nothing less than to possess all the world's known literature, a lofty ambition that he came very close to realising. Several centuries later, the scholar Callimachus detailed and catalogued some 400,000 'mixed' scrolls and 90,000 'unmixed' scrolls, these were joined by a further 42,000 housed in the Serapeum, which was a truly staggering amount of works. Amazing attempts to add to the library were carried out by Ptolemy III, who wrote a letter to 'all the world's sovereigns' in which he cordially asked if he could 'borrow all their books'. One story goes that Athens lent him the texts of Euripides, Aeschylus and Sophocles, which he had copied, kept the originals and sent back the copies. Other stories relate how all ships that docked in Alexandria were searched for books, which were then given the same treatment (this gave rise to the term 'ships libraries' in a museum). Unfortunately for Demitrius, he was never known as a librarian, dying before the Serapeum was made and seemingly not known as a librarian of the museum. The first recorded librarian was one Zenodotus of Ephesus, holding the position until around 245 BC until he was succeeded by probably the most famous of the librarians at Alexandria, Callimachus of Cyrene. This astute scholar created the first subject catalogue of the library as we have mentioned above, cataloguing some 120,000 scrolls within the library's holdings, these were called the Pinakes, or Tables. Callimachus seems to have been replaced by his much younger rival, Apollonius of Rhodes, who wrote the wonderfully meticulous epic, 'Argonautica'. He in turn was replaced in 235 BC by the famous Eratosthenes of Cyrene, a stoic geographer, scholar and mathematician who managed to write his 'The Scheme of the Great Bookshelves'. In 195 BC, a certain Homeric scholar called Aristophanes was installed. He updated the Pinakes of Callimachus, but nothing much is known about him. The final recorded librarian was Aristarchus of Samothrace, an astronomer who succeeded to the post in 180 BC, but was driven from it by continuing dynastic struggles between the Ptolemys. From this time onward in the history of the library, no other librarians are listed and scholars are simply referred to as Alexandrians.

The library did not actually have a perfectly systematic organisation, but rather tended to house new chests and shelves of papyri in the groups in which they were acquired. The Alexandrians from Callimachus onwards tried to keep track of their holdings via a subject catalogue. This catalogue followed the precedent of Aristotle's divisions of knowledge. Aristotle had arranged a new grouping of subjects that had once been universally known as 'philosophy', he subdivided these subjects into observational and deductive sciences. Here we will take a brief look at some of the subject categories that were studied to great effect at Alexandria.

Astronomy

Many Greek scientists classified astronomy as the projection of three-dimensional geometry into a fourth dimension, namely time. The movements of the stars were studied closely as was the moon and sun. These observations were essential for terrestrial positioning, since they provided universal points of reference, and this was especially needed in Egypt to settle the question of property rights when the yearly inundation often altered the physical landmarks and boundaries between fields and towns. In Alexandria, the sea port par excellence of the ancient world, the study of astronomy allowed sailors to travel safely and meant that they could do away with the time honoured practice of the study of oracles. It also meant that year round navigation, away from site of the coastline was now a viable option. Greek astronomers had mainly concentrated on a theoretical model of the universe, now the Alexandrians took up the task of detailed observations and mathematical systems. With these studies they backed up and developed the existing systems.

Schemes of the Universe

Aristarchus used Alexandrian trigonometry to estimate the distances to and size of the sun and moon. He also put forward a theory of a heliocentric universe. This angered fellow library scholar, Cleanthus, who stoically accused the poor Aristarchus of impiety. During the reign of Ptolemy VII, Hipparchus re-discovered and measured the precession of the equinoxes, the size and trajectory of the sun and the path of the moon. Some three hundred years later, the great geographer, Ptolemy (no royal relation) mathematically worked out his brilliant system of epicycles to support the geocentric view as proposed by Aristotle. He also wrote a treatise on astrology. Both of these works were to become a mainstay of the medieval world and a paradigm that would last some considerable time.

Mathematics

Most Alexandrian mathematicians concerned themselves with geometry, but we do know of some researches carried out into number theory here. Eratosthenes the librarian was a bit of a number dabbler and is reported to have invented the 'sieve'; a method for finding new numbers, the great Euclid was also supposed to have studied this subject with some success. Eudoxis of Cnidus, a pupil of Euclid, worked out of Alexandria and developed his method of integration as well as studying the uses of proportions for problem solving. He also contributed many formulas for the measurement of three-dimensional figures.

Theon and his famous daughter Hypatia also worked with astronomy, geometry and mathematics, but sadly none of their works survive them. Pappus, a brilliant fourth century AD scholar was one of the very last great Greek mathematicians; he concentrated on large numbers and constructions in semicircles. Pappus was an important link between European culture and astrology gleaned from eastern sources.

Geometry

History tells us that the third librarian of Alexandria, Eratosthenes, calculated the circumference of the Earth to within 1%. He based this on his measure from Aswan in upper Egypt to Alexandria, the fraction of the whole arc determined by differing shadow lengths at noon in the two locations. His other suggestions were that the seas were connected and that Africa may be circumnavigated, he also suggested that 'India could be reached by sailing westward from Spain'. This brilliant geometer also deduced the length of the year to 365 1/4 days and was the first to suggest adding a leap year every four years. The Alexandrians compiled and set down many of geometric principles of earlier Greek mathematicians. It would seem they also had access to older Babylonian and Egyptian knowledge on the subject of geometry. The library excelled in this area. It was said that Euclid was invited to Alexandria, his 'Elements' are seen as the foundation of the study of geometry for many centuries. Hipparchus and Apollonius carried on his work in later times and the great Archimedes also. It was he who is credited with the discovery (or re-discovery at least) of Pi. So they destroyed the ancient knowledge at Alexandria and thought it would not survive, but they had forgotten about Solomon.

Solomon was as a descendant of Moses who came out of Egypt during the Exodus. Some say that Moses was Akhenaten and there is no doubt that he would have had the knowledge of the House of Amen as a Pharaoh. Solomon built a temple to house the Ark of the Covenant in the Holy City of Jerusalem and in it he placed great underground places to house all sorts of knowledge and treasures. This Temple of Solomon is the one that The Knights Templar were excavating and I believe that the treasures that they discovered were of such enormous importance that they named themselves after the Temple. Near Roslyn is another church and village called Temple and shows the importance that the Knights Templar placed upon the discoveries they found. So despite the forces of chaos constantly destroying this kind of knowledge, it pops its head up again somewhere down the Golden Thread of Time and is resurrected in its archetypal form. The cross has been resurrected again in the twenty first century

AD to explain the mathematical skills of our ancient ancestors. But, along with the cross in its resurrection comes something else, and that something is mathematics. Let us look at the evidence with our own eyes as opposed to the historical version and go back much further in time than the Greeks who conquered Egypt and claimed the mathematical knowledge as their own.

Mechanics

One famous Alexandria affiliated scholar was Archimedes; he applied Geometers and astronomers theories of motion to mechanical devices. He discovered the lever and the so-called 'Archimedes screw'; a hand cranked device for lifting water. His famous cry of 'Eureka!' is known the world over. Hydraulics was a science born in Alexandria and this was the principle behind Hero's 'Pneumatics', a very long work that details machines and 'robots' mimicking human actions. His inventions included a windmill driven pipe organ, a steam boiler (which the Romans later adapted for the heating of their baths), a self trimming lamp and the so called candelaria, in which the heat of candle flames caused a hoop from which were suspended small figures, to spin. Children the world over should be forever thankful.

Medicine

Many Alexandrians studied anatomy, formally the study of Aristotle. The zoological gardens were probably used for animal case studies as well as various Egyptian sites where bodies could easily be extracted for human anatomy. Herophilus, who embarked on studies of his own, compiled the Hippocratic corpus in Alexandria. He first distinguished the brain and nervous system as a single unit; he ordered the function of the heart, the circulation of blood and several other anatomical systems. Eristratos who concentrated on the digestive system and effects of nutrition succeeded him; he postulated that nutrition as well as nerves and brain influenced mental disorders. In the second century AD, Galen drew upon Alexandria's vast researches and his own discoveries to compile some fifteen books on anatomy and the art of medicine.

The Museum of Alexandria was founded at a unique place and time, which allowed its scholars to draw on the deductive techniques of Aristotle and Greek thought, in order to apply these methods to the knowledge of Greece, Egypt, Macedonia, Babylon, and beyond. The location of Alexandria as a centre of trade, and in particular as the major exporter of writing material, offered vast opportunities for the amassing of information from different cultures and schools of thought. Its scholars' deliberate efforts to compile and critically analyse the knowledge of their day allowed for the first systematic, long-term research by dedicated specialists in the new fields of science suggested by Aristotle and Callimachus.

The books were used as fuel to heat the city's 4000 public baths for six months. This was the ultimate evil that could possibly have been bestowed upon this library and its custodians. Ultimately, these monstrous acts by the different religions of the time, all of whom believed they were right, have filtered down the ages to affect us and deny us the truth of our past inheritance. Some of the knowledge was known to have escaped the burning of Alexandria and had been kept at Constantinople. The Muslim Turks were known to have generated a great civilisation, which was deeply involved with astronomy and navigation. Constantinople was at the heart of their Empire and so they would have had some access to the knowledge of the ancient Egyptians. There are clues to the cross and plumbline, which can be seen in the Piri Reis map, and it is known that The Knights Templars had communications with the Islamic peoples from time to time. Admiral Piri Reis was Turkish, so it is not unthinkable that he may have acquired the maps from the Templars or vice versa.

An object of particular interest to this story can be seen on the Piri Reis maps. This object is a scale with attachments at either end.

The Pirie Reis Map

This embellishment is a good representation of the exponential scale on the cross that I have discovered in the Great Pyramid of Khufu. Please note that the knowledge of an exponential scale is clearly seen with the rose lines bisecting the scale perfectly. Furthermore, these scale rules are also represented as angled on the charts with attachments at each end for fixing to the cross bar.[xciii] This vellum chart is considered to be ancient and that the cartographers copied it faithfully from much earlier ones. It is consider to be so old that there is continued debate over the Antarctic ice caps as the chart clearly shows the east coast of the Americas with considerable accuracy and goes on to show the Antarctic free of ice. Such accurate surveys have been debated as having been carried out from the air in ancient times. The cross and plumbline and geometry are capable of producing such results. The Knights Templar did not rely on the diffused knowledge from Constantinople, as this was not the archetype. No, they found the knowledge in a more original form in the Temple of Solomon, which had kept it safe from the rampaging Romans and the burning of Alexandria. With this knowledge, charts and instructions, and armed with the cross as a navigational instrument, I believe that the Templars made forays to Mexico and other areas for over two hundred years, trading with the natives in the same way that the Phoenicians had done. This created sufficient wealth for the Templars to enable them to become moneylenders to various Heads of State but their money was more likely to be their undoing, rather than any religious objection that the other religions may have had towards them.

THE CROSS AS A COMPASS

The ability of the cross to be used as a compass unlocks secrets that have not been made public knowledge for millennia. The earliest chart to illustrate this technique is the Pisan chart of c.1290.[xciv] Hidden in this chart is a quadrant and plumbline. The lettering at the foot of the chart deliberately obscures the plum bob and if you were unaware of the knowledge of the cross and plumbline you would not understand its full implications. The potential to carry out sophisticated navigation techniques made the cross a powerful weapon of war in its time. I say the true cross, not a basic quadrant, as the missing ingredient on a quadrant is the cross bar and upright. The upright above the cross bar is essential for sidereal observation, an ability that is unique to the cross. It allowed the seafarers to make maps and charts enabling them to traverse the seas in relative safety and navigational confidence. Previously, the cartographers had only known the knowledge and techniques required for this art. More importantly though, it enabled networks of Rhumb lines to be plotted. Imagine the consequences, if the Templar battle ships were able to traverse on a diagonal route in relation to the course of a merchant or enemy ship. Then the ship using the great circle or Rhumb line method of navigation would be able to cut off the other ship, if the other ship was using the traditional methods of the times by following a line of latitude.

The hand held cross that is shown in the plate measuring the summer solstice may be used in a different fashion to the other methods that I have described. It may be laid on its back and triangulation methods applied. To triangulate with the cross is very simple, you sight the pole star to find true north and then spin the wheel west or east of the pole by the amount of degrees required to set a course at the correct angle. I think that this is exactly what happened and this ability led to the stories and legends of Pirates on the Spanish Maine.

Was the discovery of America the Greatest Confidence Trick of the Last Millennium?

Is it possible that the vengeful fleet of the Knights Templar were able to do just this to the Roman Catholic Spanish ships returning laden with wealth from the plundered Mayan, Aztec and Inca nations? To be able to triangulate meant that you did not have to follow the line of latitude, but that you could shorten the distance by following a great circle route. A great circle route is the shortest distance between two points on a globe and it requires cutting across the grid of latitude and longitude at a specific angle. It is possible to watch a ship leave port, leave after it has sailed and arrive before it at a suitable ambush point using triangulation and the great circle route. I believe that Christopher Columbus already knew about the existence of the Americas and deliberately set the Roman Catholic Spanish up for the greatest con trick in the last Millennium. Did the surviving fleet of Knights Templar, flying the skull and crossbones, manage to cut these unsuspecting Spanish ships off using the great circle route and raid their cargo? Were the Templars instrumental in the destruction of the Maya by accident or design? Did the Roman Catholics manage to torture enough information from the Templars to know about America? Christopher Columbus' sails did display the Templar cross after all and many paintings show him landing on the shores of the West Indies holding a very large cross. The black flag showing the skull and crossbones was the Templar battle flag and is still a masonic symbol. This notorious flag was also the ensign flown by the pirates that plagued the Spanish ships returning with their ill-gotten booty from the Americas.*xcv* Is there evidence of this suggestion that Columbus was working a clever ploy on behalf of the Templars and acting as a double agent? Let us see from the following information.

Did Columbus Sail the Ocean blue before 1492?

On the Piri Reis map, it's claimed that Christopher Columbus discovered the Americas in 1485/86. Obviously, historians as an error derided this claim, or simply ignored it, however, now an Italian historian and researcher claims to have found evidence that the Piri Reis claim could well be true. Ruggero Marino has found evidence, in the form of an inscription on the tomb of Pope Innocent VIII, who died before Columbus' voyage of 1492, to support the version of events as described on the Piri Reis map.

There are many other claims associated with Piri Reis that are ignored or dismissed by historians. He claimed that Columbus was using a book of charts, describing America when he made his voyage. Could this book have been a Templar book? Apparently, again according to Piri Reis, it was this book that Columbus used in order to secure the vital funding for his voyage. Again, this is simply dismissed by scholars and writers on the subject. The Piri Reis map itself is supposed to be based upon much earlier maps and charts, in fact Piri Reis claimed these went back as far as Alexander the Great. Below is reproduced the article from the London Times of Tuesday May 8th, in which the claims of Marino were first brought to light.

Of course this sheds light on something I had been wondering about for some time now. In his book about his father and based upon the great sailor's diaries, Fernando Columbus relates a tail that I will paraphrase thus. It seems that when the Santa Maria was in the area known as the Sargasso Sea and the crew were getting mutinously restless, a shout was heard from a lookout proclaiming land in the north, much to the excitement of the crew. With food and water running low, you would assume that any glimmer of hope for land would be seized upon immediately. In fact Columbus does the opposite. He calmly assures the gathered crew that there is no land to the north and that the course they are on is the correct one. Of course this amazing state of affairs becomes all the more explainable if Columbus had indeed been on this voyage before - he simply knew where he was going.

The Times Tuesday May 08 2001
 Columbus 'discovered America earlier'
BY RICHARD OWEN

AMERICA was not discovered in 1492 by Christopher Columbus, on behalf of Spain, because he had sailed to the New World seven years earlier on a secret mission for the Pope, according to scholars.

Ruggero Marino, a writer and historian, said the 1492 journey was a return visit. He said this emerged from study of an early 16th-century Ottoman map, which showed that Columbus found America in 1485, during the reign of Pope Innocent VIII.

Signor Marino said there was corroborative proof in an inscription on the tomb of Innocent VIII, in St Peter's Basilica, which reads "Novi orbis suo aevo inventi gloria", meaning that during his pontificate "the glory of the discovery of the New World" took place.

Innocent VIII died at the end of July 1492, before Columbus set sail and three months before he landed at the Bahamas. "The inscription either anticipates Columbus' success or else refers to an earlier journey," Signor Marino said.

The accepted version is that King Ferdinand and Queen Isabella dispatched Columbus in 1492, but Signor Marino said that Innocent VIII and the Medici banking dynasty to which the Pope was related originally financed the venture.

The aim was to use the gold of the New World to fund Crusades and to gain souls for Christianity. Innocent VIII's successor, Alexander VI, the Spanish-born Borgia Pope, "signed over the rights to the New World" to the Spanish throne. The origins of the New World venture had been covered up partly because his Spanish masters manipulated Columbus' writings.

Signor Marino told Oggi magazine that he had drawn on the work of the late Professor Bausani, head of Islamic Studies at the University of Venice, who had studied the Piri Reis map, named for Piri Ibn Hadji Mehmed, admiral of the Ottoman fleet. The map, dated 1513, was lost for centuries; it is now in the Topkapi Museum in Istanbul.

Professor Bausani said that the "key to the mystery of Columbus and the Indies" lies in an annotation of the map, which refers to the American land mass: "These shores were discovered in the year 890 of the Arab era by the infidel from Genoa." Genoa was Columbus's birthplace and 890 correspond to 1485-86.

It seems from the evidence that the Knights Templar resurrected ancient Neolithic sailing routes and harbours. Most of these locations also have stone circles and megaliths and this can be easily demonstrated. Since I have already shown you that these sites may be used as clocks for finding local time, it makes sense that the Knights Templar also used them for the same purpose. In the book "The Templars Secret Island" by Erling Haagensen and Henry Lincoln,[xcvi] there is an island in the Baltic which is called Bornholm which does not get a lot of publicity like Roslyn Chapel does. Yet, this amazing island is an ancient stronghold of the Knights Templar and it is also covered in megalithic giant stones. Haagensen and Lincoln also go on to show that the island was connected with the Merovingian Kings, who were also from Burgundy in France and are reputed to be directly descended from the Davidic bloodline. They show an association between the name Bornholm and Burgundy. Burgundy was the seat of the Merovingian dynasty and also the middle name of Jacques Burgundius de Molay.

The Merovingian Kings were at the height of their power between 400 and 600 AD and were known to have long blonde hair, beards and blue eyes.*xcvii* Does this ring any bells with the descriptions of the Maya and Aztec gods and teachers that their legends say brought them knowledge and civilisation? Bornholm displays remarkable evidence of triangulation and the use of the ancient Egyptian and English measure of the foot and yard. I believe that the Knights Templar not only found the cross and plumbline but also the old charts of the Earth showing the original Neolithic clocks. They also found mathematical information that allowed them to create the great Cathedrals in Mediaeval Europe and to go on to celebrate their skills by building churches on the island of Bornholm in a superb display of measurement and triangulation. This placing of churches on the island encompasses mathematics hundreds of years before its time. There is another site, not yet very well known, which is also a Templar stronghold and strategic centre. This place is on the west of Scotland and is known as Kilmartin.

KILMARTIN

Kilmartin is a little known area on the West Coast of Scotland, just north of Crinan near Lochgilphead. It is a fascinating ancient site, full of Neolithic stone circles and burial mounds and they recently opened a visitor centre displaying the history of the site and various finds. It is an area where stone balls have been found. These balls are unexplained and have grooves carved around them. I suggest that these balls were used as plumb bobs for the purposes that I have earlier outlined, as their size is similar to the one found in the Queen's shaft of the Great Pyramid by Waynman Dixon. I believe that this site in Scotland is part of the ancient transatlantic route.

It is generally known that this was the landing point of the Scoti, after which Scotland is named. This tribe originally came from Ireland and some researchers believe that the name of the tribe is associated with one of Akhenaten's daughters whose name was Scoti. I believe that the Knights Templar were aware of this ancient landing point, having discovered documen-tation pointing to it from their discoveries in the Temple Mount in Jerusalem. I also believe that they used Kilmartin as a strategic base just as the Neolithic seafarers had done before them. Temple Wood (aptly named in this scenario) in Kilmartin is home to a stone circle which is aligned to the four cardinal points of north, south, east and west. It can be seen that the north stone has a spiral carved into its face. Curiously, there is a line extending round to the eastern side of the stone. This spiral is aligned directly to the Corryvrekan, which is known to locals as a notorious and dangerous whirlpool to the west of Kilmartin. Again, we see the use of spirals all over the ancient world and on both sides of the Atlantic, surely demonstrating that the two sides of the world had some sort of intercontinental communications between them.

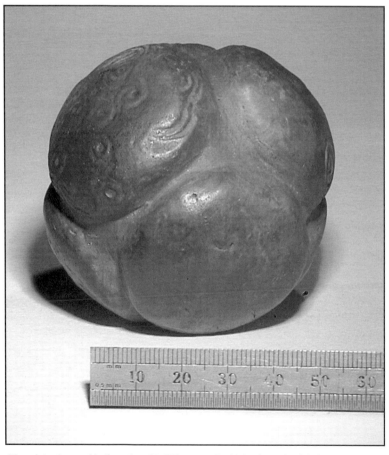

Unexplained carved balls as found in Kilmartin. Could this be a plumb bob?

Knights Templar in Kilmartin

In the village of Kilmartin, there is a graveyard, as can be found in most villages. The photograph above is a view from Kilmartin graveyard, due south across fields containing Neolithic burial mounds. To the right is Temple Wood, an ancient stone circle used for astronomy and time keeping. However, in this particular graveyard, preserved for all to see, there are a number of Templar gravestones and carvings, an example of which can be seen in the photograph on the right. In the coat of arms on the gravestone you will see the engraving of a sailing ship, and below the coat of arms, there are also the skull and crossbones as is also shown on the pirate flags of old.

Most of the tombs and carvings at Kilmartin date back to the 14th century and most of them have evidence of skulls and crossbones on the gravestones as does the tomb of William St Claire in Roslyn Chapel. Kilmartin was obviously a Templar and Pirate stronghold in its day and a very useful refuge from the Roman Catholic countries that were all over Europe at the time.

The other icon that may be seen at Kilmartin is the Fleur De Lys, which is well known as an ancient royal and religious symbol. This symbol is shown on early charts and is mounted on a compass rose to designate north. Early charts used the directions of the winds as a rose and not degrees.[xcviii] The Fleur de Lys is similar to the bronze fork on the cross and plumbline found in the Great Pyramid of Giza, except that an extra point has been incorporated. The function in navigational terms remains the same, if mounted on a cross and used to take angular, sidereal sightings of stars and constellations, the pointer would be highly efficient. Using this instrument, it is possible to find a compass bearing, provided you know the position of the stars at the time of the observation. For example if a known star is due south and its declination is first found, the next star along the ecliptic, for example, is known to be a certain number of degrees to the east or west of the southerly star. A course can then be set for that star cutting at a tangent between the longitude and latitude grid.

As I have already said, this is in fact a more accurate system than a compass, as compasses are subject to deviation and magnetic anomalies. The legend of Henry St Claire sailing to America before Columbus makes more sense now, when you consider that the natural harbour of Crinan near Kilmartin would be a perfect place for his ship to land or depart. The stone circles near the coast would give him local time in relation to Giza so that he would be able to find longitude. Remember, in the time of Henry St Claire, Greenwich did not exist, but the legacy of the ancient prehistoric mariners did and still does to this day.

CHAPTER 9:

OVER THERE

"THE TUCSON LEAD CROSSES, FACT OR FANTASY?"

Special Collections, University of Arizona Library, March - April 1990.

The discovery in 1924 of a series of metal objects, some shaped like crosses, swords, javelins, batons and paddles at an old lime kiln site, seven miles north of Tucson along Silverbell Road, ignited a controversy that lingers on today.

The "Lead Crosses,' as the artefacts came to be called were covered with Christian crosses, Moslem crescents, Hebraic seven-branched candle sticks, Freemasonry symbols, Latin phrases and Hebrew phrases. A translation of the inscriptions was interpreted to tell the story of a group of Roman Jews who migrated to America, which they called "Calalus" during the last half of the eighth century. The writings also tell of a military struggle with the Toltec Indians of Mexico. Other interpretations attributed them to the early Spanish explorers of the Southwest and to the Mormons.

A feature story in the "Arizona Daily Star" of December 13, 1925 and in the New York Times started an editorial battle of words across the country. Many prominent University of Arizona professors were quoted in the newspaper articles.

Charles E. Manier first unearthed the objects in September of 1924, at an old ruined limekiln site on Silverbell Road, seven miles north of Tucson on the property of Thomas W. Bent. The feature story in the Arizona Daily Star included photographic reproductions of the artefacts and interpretative translations of the inscriptions by a Mrs Ostrander, a local history teacher at Tucson High School. The descriptions of their discovery and unearthing were witnessed and attested to by Frank Fowler, professor of Classical Literature at the University of Arizona, Dr A E Douglass, Director of the Steward Observatory, Dean Byron Cummings of the Arizona State Museum, and Charles T Vorhies of the Agriculture College at the University of Arizona.

The New York Times reported the discovery of the lead crosses on the same date, but cast doubt on their authenticity by quoting Bashford Dean, Curator of Arms and Armour at the Metropolitan Museum of Art, who called them the "poorest of forgeries". F.W. Hodge of the Southwest Museum thought that an archaeological faker who had been making false inscriptions of the Coronado and Fray Marcos de Niza expeditions in Arizona was responsible for the Tucson lead crosses. Dr Neil M Judd, anthropologist and curator of American Archaeology of the United States National Museum, came to Tucson to examine the site and stated that he thought the artefacts were no older than 1540. Widespread interest in the artefacts continued as a newspaper battle raged across the country.

In January of 1925, under the direction of Dean Cummings, the University of Arizona took over the excavation of the site. In July of 1925, "The Improvement Era", an official publication of the Mormon Church published an article on the artefacts by J M Sjodahl, former editor of the "Desert News", who stated that "the inscriptions on the crosses were exceedingly significant.

When the "Lead Crosses" were first found, a local Tucson teacher, Laura Ostrander, translated the inscriptions. Here is some of the text of that translation. It tells a very strange and interesting story.*xcix*

The Tucson Lead Crosses, Inscription Translation

On the cross arm at the left is a profile of a head with the words *"Britain, Albion, Jacob"*. In the centre is another head profile with the words *"Romans, Actim, Theodore"*. On the right is another head profile with the words *"Gaul, Seine, Israel"*. On the vertical beam of the lead cross is this inscription: *"Counsels of great cities together with seven hundred soldiers AD 800, Jan. 1." "We are borne over the sea to Calalus, an unknown land where Toltezus Silvanus ruled far and wide over a people. Theodore transferred his troops to the foot of the city Rhoda and more than seven hundred were captured. No gold is taken away. Theodore, a man of great courage, rules for fourteen years. Jacob rules for six. With the help of God, nothing has to be feared. In the name of Israel, OL."*

The second cross has the following inscription: *"Jacob renews the city. With God's help Jacob rules with mighty hand in the manner of his ancestors. Sing to the Lord. May his fame live forever. OL."*

The third cross yielded this inscription: *"From the egg (the beginning) AD 700 to AD 900. Nothing but the cross. While the war was raging, Israel died. Pray for the soul of Israel. May the Earth lie light on thee. He adds glory to ancestral glory. Israel, defender of the faith. Israel reigns sixty-seven years."*

The next inscription: *"Israel II rules for six. Israel III was twenty-six years old when he began to rule. Internecine war. To conquer or die. He flourishes in ancestral honour day by day."*

The next inscription: *"AD 880. Israel III, for liberating the Toltezus, was banished. He was first to break the custom. The Earth shook. Fear overwhelmed the hearts of men in the third year after he had fled. They betook themselves into the city and kept themselves within their walls. A dead man Thou shall neither bury nor burn in the city. Before the city a plain was extending. Hills rung the city. It is a hundred years since Jacob was king. Jacob stationed himself in the front line. He anticipated everything. He fought much himself. Often smote the enemy. Israel turned his attention to the appointment of priests. We have life, a people widely ruling. OL."*

The next inscription: *"AD 895. An unknown land. Would that I might accomplish my task to serve the king. It is uncertain how long life will continue. There are many things, which can be said while the war rages. Three thousand were killed. The leader with his principal men is captured. Nothing but peace was sought. God ordains all things. OL."*

Here is what Jack Andrews[c] has to say:

"The text following this introduction is from a small introductory paper written for an excellent display that was organised by Peter L. Steere of the Special Collections staff at the University of Arizona Library. No author is listed, but I give full credit to Special Collections for the excellent and informative article. I post it here for educational purposes and research. I do not necessarily agree with some of the statements in the article, by various people. I have discussed the subject of caliche formation with Dr. Cyclone Covey (mentioned in the article) and a few Geologists. From my research, the subject of the rate of deposit of caliche is still somewhat of a mystery. Some of the artefacts were buried under at least five feet of hard caliche. Dr. Covey has said in a communication to me: "Concerted efforts to discredit (the age of the find) have found no way to insert heavy lead objects up to 6 1/2 feet deep through caliche to lie flat, without fracturing the formation conspicuously"

communication March 26, 1999

I agree here with Dr. Covey. In the early days in Tucson, caliche deposits sometimes called "desert cement" were a constant problem to those who needed to dig in the soil. (As they are today) Some of the "old timers" resorted to dynamite to break through the thick hard layers. I have personally used a jackhammer to dig holes large enough to plant small trees. (When I worked as a landscaper) here in Tucson. I find it hard to imagine that Timoteo Odohui buried 30+ heavy lead objects nearly six feet deep through solid caliche by himself, and then

covered all the objects, simply to create a hoax. When the objects were excavated with University of Arizona supervision, It took a crew of several Mexican
labourers using picks to forcefully hack their way down to the objects through
the hard cement-like caliche. Also, some of the objects were personally witnessed to have been dug from undisturbed caliche and soil, by University of
Arizona professors."

Jack Andrews
Jack Andrews copyright 2001 may be only used in its entirety with this notice clearly visible.

Of course, this lends credence to the claims by the Mormons of early settlement
in America. In conclusion it seems to me that mariners have been crossing the
Atlantic since the Pleistocene age at various times in the history of the planet and
mariners were experts in astronomy so as to tell their position through time
keeping and observation. This knowledge, known as archeo-astronomy, became
sacred and was incorporated into the building of the mysterious ancient edifices
still to be seen around the globe. That there was a development from simple
astronomy to a high form of art incorporating astrology and that this leads to a
form of spirituality that has since waned in the world. It also seems on weighing
the evidence, that there is a strong possibility that there were cyclical events that
caused these eras of voyaging to be terminated and then rediscovered. In later
times from Ancient Egypt onwards, that these events were mostly caused by
human turmoil, division and war. The knowledge of the cross as a mariner's
instrument became secret to only the enlightened few. These were probably the
ancient Neolithic mariners, followed by Phoenicians. The next to use the instrument were very likely the fleeing Essenes who influenced the Hopi and Mayan
people and lastly the Knights Templar a thousand years afterwards.

Did the Romans deliberately crucify Christ on a replica of an instrument of great
power and wisdom?

Were the Neolithic astro-astrologers able to use the cross to align their stone
circles to the equinoxes and solstices?

Could the Greeks have had access to it at Alexandria and was the knowledge lost
through the criminal burning of the Library of Alexandria?

Was the cross found again by the Knights Templar during the Crusades in
Solomon's Temple and used to open ancient trading routes around the world?

Did the Knights Templar keep the knowledge in secret in Scotland and Portugal until this day or did the Roman Catholic Church extract this knowledge from the Knights Templar through torture and keep the knowledge a secret until now? Was the cross used secretly in the great Cathedral building age in Europe and is that the hidden message where artists show saints kneeling before its image? Was the opening of the exploration of the seas by the Portuguese, influenced by the Knights Templar or Knights of St John? Did the Knights Templar wreak revenge on the Spain and other Roman Catholic countries, by committing acts of piracy on the high seas under the flag and symbol of the Knights Templar's skull and crossbones? Was the use of the cross suppressed in Spain and other areas of Europe under the Spanish Inquisition through ignorance or because they did not want the secret of it to be revealed? Was the working cross destroyed by the Conquistadors in Mexico, The Yucatan Peninsular and the Andes along with the societies that used it? Did the Conquistadors attempt to annihilate the Maya, The Aztec and the Inca peoples, because they thought the use of the cross to be heresy, or because they wanted to hide the secret? Was human sacrificing an excuse or propaganda for their cause? There are so many unanswered questions.

I believe that our ancestors taught the Mayan People the secrets of the cross and the mathematics exactly as their legends describe, and I also believe that several hundred years later, the descendants of those benefactors returned and destroyed these people. It is fair to say that some of the Amerindian people misconstrued the knowledge that was given to them. Whereas they used the mathematics, astronomy, agriculture and building methods to a level of supreme expertise, they did not understand the meaning of sacrifice by giving one's heart to God. They considered this concept in a selfish and ignorant way and performed human sacrifice by removing other people's physical hearts to appease the spirits of nature, hoping to protect themselves from cyclical disasters. This horror and ignorance did not justify the slaughter and destruction that befell them at the hands of the Conquistador. I have apologised on behalf of my ancestors to the Mayan people before the end of the last millennium and it happened like this.

A Step Towards Reconciliation?

I am quite active on discussion groups on the Internet, and many had heard of my discovery of the working cross. In early 1999, Olympia Xux Ek who was organising a conference in Tenerife through her Mayan Mystery School approached me. The Conference was called the conference of Chakuna and was scheduled for April 30th 1999. It was to be held at a beautiful hotel called the Hotel Melia at Puerto De La Cruz.

At this stage, I had no idea there were Mayan and Inca Elders attending the conference. I also had very little knowledge of the Mayan history or calendar. I set about designing a cross that would be easy to assemble and would break down to travel on an aircraft. To achieve this, I went to my local DIY store to purchase all the necessary materials to construct the cross. I needed to construct an instrument where the cross bar would rotate so as to fold into one length. I purchased two lengths of half round rope styled architrave of two metres in length, a nut, washer and bolt for the central swivel, a one meter wooden rule, a plumbline, a plumb bob and a gold kitchen knob.

When I arrived back at my workshop, I commenced assembly of the instrument. I cut the appropriate length for the cross bar from one of the lengths of architrave, glueing and screwing the remaining part to the bottom of the remaining full length. I fixed the centre of the cross bar to the upright pole with the bolt, so that the cross bar could swivel out at right angles to the upright. The whole assembly would rotate on the swivel so as to look like one length of roped pole or staff when it was closed. I decorated the end of the bolt with the flat gold kitchen knob, so that it would be seen at the centre of the cross when opened or shut. To hold the staff together, I looked for a piece of rope to pull through a hole that I had drilled at the top of the staff. I eventually found an old piece of my daughter's boat's rigging that would work. This was a multi-coloured length with a knot and tassel at the end. I cut the ruler to 90 centimetres and affixed that to the assembly in such a way as to easily remove it for travel. The whole assembly was then placed in a fishing pole tube for transportation.

I also decided to take a series of slides with me, demonstrating how the instrument could be used.

The whole trip was amazing. I met some wonderful people, and took part in several Mayan ceremonies. We took part in a sun ceremony, where the Mayan priests dug a small round pit in the sand. On the bottom of the pit they laid out a perfect Celtic cross design using sugar. Between each cross arm was a small circle to represent the four winds. The Mayan priests were totally aware of the four cardinal points. A fire was built up using pinecones and sweet smelling spices. The top was decorated with coloured candles, ranging from yellow in the east cardinal point through orange, green, blue and black in the west. This is to symbolise the day and life as the sun travels to the west, and life moves from its beginnings to the end.

Later that day we visited the Ethnographic Park of the Pyramids in Guimar, where there is a museum showing Neolithic carvings and stones. Many of these carvings show spirals in the same fashion as those Neolithic designs found all over the Atlantic seaboard from the Orkneys to Brittany and Ohio in America. The authorities at the time were still claiming that the pyramids were a result of the farmers clearing land. This is a highly unlikely explanation because of the precise orientation of these edifices. They are very small when compared with the pyramids of Giza and the Yucatan. They are however, seven step pyramids and are precisely aligned.

The one we visited is on the east coast of Tenerife, but there are also remnants of pyramids on the west coast. Furthermore there are pyramids on the east and west coasts of Las Palmas. It was here that I heard there was a Mayan style carving of a face that had just been found in Las Palmas, the most western island of the Canaries. The museum has replicas of both Thor Heyderhal's Ra and a large model of Christopher Columbus' ship. The guides have a policy of allowing visitors to look and make up their own minds about whether or not these pyramids are indeed true pyramids.

They are very protective of the site and do not allow people to walk on the pyramids. In this case however, we were allowed to spend an hour there and observe the site at close quarters. The Mayans had made a request of the authorities that they be permitted to hold a ceremony on the pyramids and were allowed to conduct a small one amid considerable nervousness on the part of the guardians. This could have been the first time in hundreds of years that a Mayan priest had stood on these islands and conducted sacred ceremonies to Nature, the Sun and the Great Creator Spirit. In my opinion however, these stone edifices were designed as a clock based on finding local time. They function as a water clock and can observe the reflection of both the sun and moon on the ocean surface. They also function as a clock at night, in that a trained observer can watch the altitude of the stars as they pass overhead every night and move south or north according to the season.

It was also clear to me that this had been the purpose of these pyramids for thousands of years. That they were used by the Neolithics, Phoenicians, Celts and latterly, the Knights Templar. Arthurian legend point clearly to this and it is curious that the neighbouring island of Lanzerote translates to Lancelot. The Knights Templar held dominion over these islands with permission of the Pope for a long time before they fell out of favour. These islands are the gateway to the Americas for sailing ships.

The name Canaries is again an interesting word game in that CAN means serpent and wisdom in the Mayan language. ARIES means the age of Aries. Combined, the name means the serpentine wisdom of Aries. Pointing to a time before Christ, when the astrological age was in Aries or Amen, again serpents raise their heads. This conference was in the name of the Rainbow Coloured Serpent or Quetzecoatl of the Ancient Mayan People. Serpents and pyramids, spirals and time keeping, ships and navigators.

I had the honour of presenting a working cross to Don Alejandro Cirilo Perez Oxlaj, President of the Mayan Elders of Guatemala and President of the Council of Mayan Elders of America. Born on February the 26th 1926 in Quetzaltenango, Guatemala, Don Ajlehandro's Mayan name is Wakatel Utiw/K'iche or in Spanish, Lobo Errante or Wandering Wolf. Don Ajlehandro is married in the Mayan tradition and has three daughters. His profession is a Mayan priest and teacher. He is the 13th in a lineage of Mayan priests. When he was thirteen years old, he was symbolically given the sacred bundle that denotes Mayan priesthood, having completed 13 cycles and completed a pilgrimage to Mayan K'iche sacred sites. He lived in Mexico for five years and studied with Huichols, Zapotecs, Tarahumaras, Yaquis, Aztecs and Otomis. He is an astronomer, astrologer, astronomer, naturopath and ecologist. He serves the community of San Francisco el Alto, Totonicapan. Since 1988 he has been a master in Mayan spirituality and has guided more than 200 disciples. He has published a Mayan calendar and continues to travel on an international basis as a teacher, priest and elder, representing Guatemalan and international indigenous organisations. Don Ajlehandro has presided over numerous events in Guatemala and in other lands. He was special guest at the inauguration ceremony of the President of Guatemala. He has lectured at Harvard University, the Sorbonne, The Egyptian Senate and the United Nations. He has also been the special guest of Yassir Arrafat and Prime Minister Yitza Rabin.

My speech lasted over two hours, and included a demonstration of the working cross. At this demonstration I showed the conference how the time could be told, how pyramids were built and how precession could be measured. At the end of my presentation I gifted the model that I had made to the Mayan people with an apology on behalf of my ancestors for the suffering and dispossession that had resulted in our peoples' actions over 500 years ago. Don Alejandro stood up and made a speech of thanks, in which he said that this instrument was his peoples' lost staff of power that had been taken away when they were dispossessed. He said that this instrument was party to the building of pyramids and glyphs showing the cross could be seen in many sites throughout the Americas. He also said that our Christ was a man not only of spirituality, but also of science, when the two were once one. He said that the powers of the time tried to destroy Christ on the symbol of his knowledge of spirituality and science. Don Alejandro then went on to say that he would present the instrument to the Council of forty five Elders and that it would be used for teaching the young in the Mayan Mystery Schools.

Earlier in this book Gary David wrote an article about the Hopi Indians and the Orion correlation. I will now tell you of the Hopi Nation's prophesies about the future.

THE HOPI PROPHESY$_{ci}$

The next war will be "a spiritual conflict with material matters. Material matters will be destroyed by spiritual beings who will remain to create one world and one nation under one power, that of the Creator." A song sung during the Wuwuchim ceremony foretells the time. The Hopi and others who were saved from the Great Flood made a sacred covenant with the Great Spirit never to turn away from him. He made a set of sacred stone tablets, called Tiponi, into which he breathed his teachings, prophecies, and warnings. Before the Great Spirit hid himself again, he placed before the leaders of the four different racial groups four different colours and sizes of corn; each was to choose which would be their food in this world. The Hopi waited until last and picked the smallest ear of corn. At this, the Great Spirit said:

"It is well done. You have obtained the real corn, for all the others are imitations in which are hidden seeds of different plants. You have shown me your intelligence; for this reason I will place in your hands these sacred stone tablets. Tiponi, symbol of power and authority over all land and life to guard, protect, and hold in trust for me until I shall return to you in a later day, for I am the First and I am the Last."

The Hopi were told that they must hold to their ancient religion and their land, though always without violence. If they succeeded, they were promised that their land would be a centre from which the True Spirit would be reawakened.

When he returns, the True White Brother will bring no religion but his own, and will bring with him the Tiponi tablets. He will bring with him two intelligent and powerful helpers, one of whom will have a sign of a swastika and the sign of the sun. The second helper will have the sign of a Celtic cross with red lines (representing female lifeblood) between the arms of the cross. The Hopi were warned that if these three beings failed, terrible evil would befall the world and great numbers of people would be killed. However, it was said that they would succeed if enough Hopi remained true to the ancient spirit of their people.

I wrote the following poem in honour of our brothers the Mayan and Hopi Nations, who struggle to maintain the knowledge known to their ancestors and pass it on through their younger generations as the world of materialism threatens to destroy their culture forever.

Indian Spirit

Where is he?
Where is the Man?
Cold blue bluff rises from the mist swathed forest.
An eastern peak capped with gold
Kissed by the morning Sun.

Where is he?
Where is the Man?
The scent of Lynx and Fox rise on gentle air.
Life stirs.
Warmed by the morning Sun.

Where is he?
Where is the Man?
The Eagle soars in silent sky to golden clouds.
Ribbon thin.
Lit by the morning Sun.

Where is he?
Where is the Man?
On the lonely bluff to dance to the New Day?
And the ancestors?
To greet Grandfather Sun.

Where is the Man?
Drifts his life away?
Numb in the stockade of ignorance and greed?
Far from Grandfather Sun?

Where is the Man?
The dust on Lonely Mountain grows
A footfall sounds in distant time.
Comes he back?
Great Spirit hear us.
That he may dance to Life and greet the morning Sun.

© Crichton E M Miller 2000

Perhaps the message from Christ that "the meek shall inherit the Earth" is one for us to think about in these times. The meek in my way of thinking are the children of the indigenous people of the planet, who are closer to nature than those of us in the sophisticated west. People who are close to nature, are far more likely to survive a cataclysm or the destruction of a sensitive and fragile economy than we are. Is this Mayan child the face of the meek to come? Do not underestimate her intelligence, her forefathers built pyramids and invented the first binary calendar in the world. She, and all the indigenous people of the Americas, are dispossessed and have been since the Christians destroyed their civilisation five hundred years ago. Look into her beautiful eyes and understand the "Golden Thread of Time" as she awaits the resurrection of the ancient spiritual and scientific knowledge of her ancestors.

Photo courtesy of Jean Parry Williams

Lost or Secret Knowledge?

It seems from the research that I have undertaken and that which many other writers have carried out, that peoples of the world in ancient times had unknown instruments and were able to use them for all the claims that I have made in this book. In the case of the remarkable mathematical discoveries of Haagensen & Lincoln and the Templar Secret Island of Bornholm, they are able to demonstrate the evidence of this knowledge in the same way that David Ritchie does in his work. Robert Bauval has also developed an insight into the evidence of the Egyptians' mathematical knowledge with his Orion Correlation, as has Gary David. Again Maurice Cotterell shows this in his work on Lord Pacal in Palenque. They have all demonstrated that the ancients were able to achieve these remarkable methods of surveying, but no one has been able to show the world how they did it. I have provided that missing key of wisdom and knowledge. Proving that the cross and plumbline is the only justifiable instrument that could make these measurements and triangulation.

However, it seems that cyclical events have taken place resulting in this knowledge being lost and recovered several times in the past. Possible causes for knowledge to be 'lost' include war and displacement, protection of trading secrets, and the maintenance of power bases. Knowledge was often lost through religious persecution in days gone by but in more modern times, maybe it has been through guilt and protectionism? After all, the grandfather of modern science, Leonardo Da Vinci, was persecuted and threatened with being burnt at the stake. He was branded a heretic and spent the last four years of his life incarcerated after being forced to confess his "Sins." His greatest sin was to claim that the Earth was not the centre of the Universe, but that it revolved around the Sun. He told the truth but the powers of the times did not like what they were hearing and thus, Da Vinci suffered greatly for it. What Leonardo da Vinci did manage to achieve was that he broke the mould and allowed the free thinkers to come out of hiding. Everything we now possess in modern times is based on their work.

We truly stand on the shoulders of Giants of courage, as well as intellect.

BRIEF HISTORIC QUESTIONS AND ISSUES RAISED ABOUT THE CROSS

- Was the distribution of Clovis arrowheads evidence of pre Flood mariners?

- Was the cross used by megalithic builders 5000 years ago in Scotland, England and France, and is the shape of Callanish as a Celtic cross in the Isle of Lewis evidence of this?

- Was the earliest Glyph in Egyptian writing, Nwit (meaning city or place), a sign that the ancients used the cross six thousand years ago?

- Did the Egyptians at the start of the age of Aries know how to use a cross and plumbline, enabling them to conduct accurate surveys of the topography and efficiently plan their buildings?

- Did the cross and plumbline allow the ancient Egyptians to develop astronomy, astrology and the use of spherical geometry leading to an accurate understanding of the nature of time?

- Was this the fundamental understanding behind their stable systems of government, based on their scientific and spiritual knowledge of the twin forces of order and chaos inherited from a maritime history?

- Did the Phoenicians and other maritime peoples use the cross as an essential instrument for navigation, allowing intercontinental travels and trades?

- Did the Israelites led by Moses or Akhenaten and members of his family take the knowledge with them when fleeing Egypt to Jerusalem, Ireland and America?

- Did the ancient craft masons of Solomon use the cross to build the temple at Jerusalem?

- Did the Essenes use it to flee from Qumran to the Antilles and then to the Yucatan establishing a new order amongst the Mayans?

- Later, were further tribes of Israel forced to flee to America from the troubles at home as shown in the evidence of the Tucson Lead Crosses?

- Were the Merovingian Kings or the Knights Templar the source of the legends of Quetzecoatl, the rainbow feathered serpent?

- Did the Knights Templar resurrect the cross and the mathematical knowledge of the ancients?

You may judge this piece of circumstantial evidence. There is some remarkable evidence that the Israelites were in North America in 800 AD shown by the discovery of the Tucson Lead Crosses.

CONCLUSION

The ancients had a knowledge that was both practical and spiritual, a knowledge that was developed from reason and common sense using both sides of the brain and taking into account both male and female as part of the whole in all things. They understood the cycles of time and our place and nature within the material and non-material world. To achieve this great insight they understood that all creation was mathematics and that mathematics helps us understand Nature. They measured the cosmos and nature with the cross and plumbline and that is why it is a revered but forgotten symbol today.

To just see it as an instrument of pain and suffering and a symbol of the resurrection is not enough for it deserves more understanding than that. It is to be revered as an instrument that encompasses the total understanding of knowledge of time, speed and distance. It was the ancient wise ones' measure of the universe and themselves and with it they maintained Ma-at, the Egyptian name for truth, balance and justice. The cross is lost in these modern times and unknown to those who have placed great faith and hope in it, and yet they cannot, and do not, see it for what it is, even though replicas of it can be seen everywhere on churches, graveyards and paintings. Those who seek understanding of the ancients by sifting through the dust of ages and measuring the constructions that were built with it, are unable to recognise it for what it is.

The cross is displayed behind the participants of a weekly religious discussion group on British television every week, viewed by millions. This is not any old cross but a perfect representation of the working Celtic cross in its full image and nobody notices it. No one, it seems, understands its hidden, often bloody, mysterious history and the part it played in the science, building, astronomy, astrology, timekeeping, navigation and cosmic understanding of our ancient ancestors.

This is a major question about ourselves, it seems we are brainwashed into not seeing things as they are in reality and that means we are blind in the Biblical sense. If we can't see the cross then we can not see each other or ourselves. We are too conditioned by society and our peers to see clearly and we must wake up, if we wish to live as sovereign individuals.

The cross was lost and forgotten till now. Why is it lost? Are the reasons religious, political or accidental? When I made and used a working model of the cross, it demonstrated itself to be a superior instrument to the quadrant by way of its ability to take sidereal observations coupled with the steadying fulcrum. I have proved, in a logical method, that this technology was available to the ancients. There is huge evidence that they used it in their mathematics, buildings, myths, exploration and writings. It is said that a plumb line and scale was found in the tomb of Rameses II, although I cannot support this with hard evidence at this time. The evidence that the ancients settled the world making stone circles and astronomical measurements as far back as Palaeolithic times is still being questioned, and it is in the political interests of many that it remains so. This can be seen by the huge academic battles over the worldwide similarity of the Clovis points mentioned earlier.

My further hypothesis on how they may have used their observatories to find longitude and keep time is still speculative as far as cautious archaeologists are concerned, but they don't have a better or more convincing argument than mine. I am sure many will say that we still require further hard evidence that the ancients travelled the oceans of the world and measured the Earth with great accuracy, despite the evidence presented of the similarity of science and artefacts on different continents and islands. But then, the medieval peoples of Europe thought the world was flat and that is why the Knights Templar and Christopher Columbus fooled them, do you think that we are any brighter today than they were? That this knowledge was rediscovered throughout history right up to the Middle Ages and maintained in secrecy is also speculative. However, they say there is no smoke without fire. The circumstantial evidence that I have presented, if held in a modern court of law, would possibly be sufficient to find in my favour. Should this book make you think and explore for yourself, then it would be a further step to winning your freedom of thought from those who by their qualifications and the prestige they covet, think they know what is good for you.

As the old saying goes "You can't please all of the people all of the time". I do not set out to please, but to shake, stir and grind to dust the illusions, lies and falsehoods of the past. I also think that "You can't fool all of the people all of the time" and that is precisely what has been happening to us for the last two thousand years or more in our religious and historical institutions.

If you believe the ancients made and used the Cross and designed the pyramids with it and that they devised the measurements of the world in degrees and nautical miles; if you see that they left the magic number of 666 in their writings and stone pyramids for us to find so that we could understand how they measured the Earth on which we live; should you agree that the many astrological and spiritual symbols written in the religious books and carved in Glyph and symbol all over the world show the meanings that I have revealed. If you believe they invented the zodiac as simulacra of the shapes in the ecliptic so that they could measure time and predict the future with a science and skill that we are only just beginning to understand; then Thoth the God of Wisdom is here again in you and your universal mind.

The madness of murder, war, division, greed and exploitation will soon be over and the dawn of a new brighter age of balanced knowledge and understanding through reason and common sense, will be emerging in the world as our ancestors promised long ago. If it does not happen in this way, then mankind is lost as a species and rightly so. It does not take the third eye to see that things cannot go on forever in the manner in which they are today.

At the start of this book, I wrote a poem and one of the lines was:

> *Order will return for a while in the cycle of the times,*
> *but the World will never die.*
> *The Vine of Man and Woman may wither and pass away*

If you think that the ancients were right and that reincarnation is a real and logical possibility, then the children of the future are your future, so teach the children to be aware of their relationship with the cosmos as your survival may depend on it. Those who exploit the world and destroy the beauty of its balance and habitat for a short lifetime of greed, power and prestige may think that they can escape through death and oblivion. Their selfishness is the cause of misery in the world that they are part of and they should remember that they may have to come back again and live in the kind of world they have made. We are made of the dust of ancient stars and driven by the energy of the sun, we are miracles of existence, part of the whole and all brothers and sisters in the world together with all creation. What we do to each other, we do to ourselves in the end, so feed the hungry, help the poor and shelter the homeless.

I am not an island and the efforts of other seekers after truth have assisted me to discover the cross and its real applications. Their hard work has enabled me to play my part in the further journey of discovery by adding my own skills and point of view to the quest for our past history. Now you know some of the answers and many others do too, which means that the ancient truth about the cross and our ancestors will continue long after we are dead and gone. I only hope that more discoveries may be added to my hypothesis to reveal the "hidden truth" and that the publication of this book may start a new treasure hunt for further evidence of our brilliant Stone Age Ancestors' underestimated abilities of both science and spirituality. If those who wish to maintain their power and authority try to suppress this awakening, they may succeed, but only for a while as the cycles of creation make the cross and its ancient, sacred knowledge reappear over and over again in The Golden Thread of Time.

APPENDIX I

THE
PATENT

Patent Application for a Survey & Navigation Device

This invention is concerned with survey and/or navigation - and is particularly, but not exclusively concerned with angular determination or measurement.

Such angular determination is required for terrestrial measurement of structural features, landscape slope, grade or pitch in surveying - and (orbital) elevation, or azimuth, of celestial bodies in navigation and astronomy.

In practice a broad division arises in angle determination between, on the one hand,

- relatively close (say, less than a mile and typically less than, say, 100 metres), stationery, or terrestrial (earth-bound), target objects, such as buildings, or immediately surrounding landscape features,

and, on the other hand,

- longer distance (i.e. global scale), (relatively) movable (i.e. in orbit), extra-terrestrial celestial bodies.

The scale of measurement, and the accuracy required, broadly dictate the measurement methodology and type of instrument deployed.

That said, some, albeit marginal, overlap arises between broad navigational and surveying requirements - thus admitting a duality of roles, for an instrument capable of angular determination and measurement.

Visual or optical wavelength sighting aside, contemporary, radio-based, systems, such as Radar (Radio Direction and Ranging) and GPS (Global Positioning Satellites), represent a convergence of navigation (from positional determination to weapons target-sighting) and surveying - allowing accuracy's of a metre or less.

Background Art

NAVIGATION

Early terrestrial and nautical navigation relied simply upon available individual, line-of-sight, 'static', reference landmarks, whether on land, or at sea.

A raw judgement of bearing and distance and intersection or triangulation of bearing points allows a primitive position fix.

Absent some reliable, relatively stationary, ground-based reference, attention turned to celestial bodes, visible to the naked eye, in orbit across the sky, at day or night - vis the sun, moon and stars.

The local time, and periodicity of motion, of planetary systems, at least in the solar system, then become considerations in nautical navigation.
The inventor has hypothesised that, as early as Ancient times (circa 1st Century AD), albeit rudimentary devices for practical angular determination had been contrived.

Subsequently, the Polynesian race - generally accredited with colonisation of the Pacific Islands - are believed to have relied upon some non-terrestrial reference measures.

On a technical and scientific front, a major advance in navigation was the introduction, in the 12th Century, of the magnetic compass into Europe, allowing direction to be determined in relation to a magnetic north pole - albeit not distance or specific location.

Moreover, magnetic direction is susceptible to local variation and distortion by proximity of metal, and is indeterminate near the poles themselves.

It was not until the 18th Century, that navigation at sea became more reliable - with the invention of:

- the chronometer (John Harrison, circa 1759), in keeping accurate time, for longitude determination;

- the sextant (John Hadley and Thomas Godfrey, circa 1830), allowing angular sighting of celestial bodies;
- and the production of British Nautical Almanac (1765).

If the time and position, of individual, or (collectively) several, celestial bodies is known, (nautical) almanacs, or sidereal (time)tables, can be used to determine location upon the Earth surface - particularly valuable upon the oceans.

SURVEYING

Generally, surveying determines and delineates position, extent and form of (proximate, stationary) features on, or beneath, the Earth's surface.

Historically, in, say, Egyptian Pyramid construction, some means of continuous terrestrial reference alignment and checking would have been required.

Fragmentary archaeological finds have supported the present Applicant's hypothesis that plumb lines and weights were used in early surveying, in order to strike a reference line - from which angular (offset) could be derived by some form of scale.

Diverse surveying instruments have since been devised, for distance, size and inclination measurement, with varying precision.

Thus, a so-called, theodolite is commonly used for direct read-out of (relative) levels, and thus indirect calculation of level change, or inclination, over a given distance or span.

Effectively, a tripod-mounted, gimbal-supported, telescope is sighted upon a remote upright scale, set at a reference position.

Provision is made for orientating the telescope sight horizontally, by reference to a spirit-level bubble adjustment, through a screw vernier.

A gimbal mount allows swivel of the levelled sight to a relative bearing, allowing triangulation from multiple relative sights at different distances and bearings.

This requires a precision engineering instrument - expensive to produce and vulnerable to damage.

Electronic or computerised theodolites, some with laser light ranging, have been devised, for data logging of multiple co-ordinates and associated elevations, in relation to (common, universally-agreed) map survey datum.

Related, construction industry, local-scale, levelling, grading and sighting reference and mutual alignment devices include:

GB2317012 - a profile board, constructed from a rigid post and moveable crosspiece, used to set levels on construction sites;

GB2298275 - a level indicator, using a pendulum, with fixed reference to determine horizontal.

GB2292220 - a levelling aid, comprising two pivotally connected arms, one arm acting as a plumb bob, employed to determine horizontal, or to set a specific angle.

Simple variants of such devices may also have a utilitarian role as, albeit primitive or crude, educational or teaching aids - in order to demonstrate raw principles, allow participation in practical application of mathematics, and in particular geometry - and in doing so, reinforce appreciation and understanding of underlying theory.

For example, a rudimentary (in)clinometer - to measure inclination, slant or slope - can be assembled from a protractor, with a string and weight attached.

By measuring an observer's distance from a target object, and using the clinometer to calculate the angle from the observer to the top of the object, basic (triangulation) geometry can be used to calculate the object's height.

However, the accuracy, consistency or repeatability and so reliability of such a (hand-held) device approach is limited.

The present invention is concerned to afford a device for robust, consistently repeatable, determination of elevation.

An overall (stark) simplicity and ruggedness of construction, stability and a degree of precision are desirable - in a single integrated device, combining functions, or roles, of survey and navigation.

Such a device could be used where resources do not allow sophistication - such as in the third world - and as a back up, should more complex systems fail.

Statement of Invention

According to one aspect of the invention, a (measurement) device comprises an elongate, unitary, member, [one end] acting as a pointer,
for sighting upon a target,
by viewing either along its length,
from the opposite end,
or from alongside,
with the other end as a support fulcrum;
a fixed scale, mounted upon the member,
a pendulum or weighted plumb line,
freely suspended from the member,
to intersect the scale,

allowing (direct) scale reading,
for determining member inclination,
and thus target elevation.

According to another aspect of the invention,
a (measurement) device
comprises
a plurality of intersecting arms,
disposed in a cross configuration,
one arm deployable as sight line,
for line-of-sight viewing longitudinally,
from one end to another,
in alignment with a remote target object;
another arm deployable as a fulcrum,
to provide stability, and
to act as pivot,
about which to adjust sight-line orientation,
inclination and/or elevation;
a pendulum or weighted plumb line,
freely suspended from the arm intersection;
and a scale, interposed between the arms,
below the suspended pendulum, or plumb line,
to allow a direct scale reading,
for determining sight-line inclination, or elevation,
of a target object.

Generally, the overall (degree of) accuracy, or angular resolution, is determined by the relative scale, or absolute physical size, of the device - a larger scale enabling a more precise determination.

For example, a linear scale rule, of some 90 centimetres in length, allows accuracy to individual minutes of arc - when deployed to bridge the cross-arms of a device according to the present invention.

A one centimetre sub-division of such a scale is equivalent to one degree of arc, in angular determination - and which, in longitude, subtends some 60 nautical miles at the Earth's surface.

Thus, 1 millimetre is equivalent to 1 minute of arc - which in longitude represents some 6 nautical miles.

However, if the device is too large, it risks becoming cumbersome and unwieldy.

A comprise, between accuracy from scale size, and accuracy through ability to operate, must be struck.

Embodiments

There now follows a description of some particular embodiments of a 'dual-role' - i.e. navigation and/or survey - instrument, for angular determination, according to the invention, by way of example, with reference to the accompanying diagrammatic and schematic drawings, in which:

Figures 1A and 1B show a (measuring) device, comprising a unitary sighting and support member, with 'fixed' [arcuate] scale and freely suspended pendulum or weighted plumb line.

Thus, more specifically:

Figure 1A shows a unitary member, disposed generally upright, or inclined to the vertical, with a suspended pendulum, or weighted plumb line, at some 90 degrees inclination (i.e. to the horizontal), as recorded upon scale subdivision markings.

Figure 1B shows the member pivoted about a lower end fulcrum, such that the distal, pointer end is aligned with a target [object].

The pendulum or plumb line intersects the arcuate scale, allowing direct reading of the inclination of the member - and which, if sighted correctly, allows the angle of elevation of the target to be derived.

Figures 2A and 2B show a (measuring) device, comprising a unitary sighting and support member, carrying a pivoted, freely-rotatable, offset bias-weighted arcuate scale - with an index on the member for taking scale readings.

Thus, more specifically:

Figure 2A shows the member disposed generally vertically, or upright, with the index, indicating 90 degrees inclination, on the rotatable scale.

Figure 2B shows the member pivoted about a lower end fulcrum, such that the upper, distal, pointer end aligns with a target [object].The member index, points to the angle of inclination on the arcuate scale - which, if sighted correctly, allows the angle of elevation of the target to be determined.

Figures 3A and 3B show a multiple (in this case dual) element (measuring) device, of general, differential-length, cross-arm configuration, with a fixed arcuate scale interposed between the arms, and a pendulum, or weighted plumb line, freely suspended from the cross-arm intersection.

Once assembled and erected (in the case of a collapsible structure), the cross-arms are set in a fixed relative disposition (in this case mutually orthogonal) - as a co-operative pair.

One (conveniently a generally longer) arm used for overall support and stability.

The other (relatively shorter) arm deployed to 'sight' - by longitudinal axial alignment with - a target.

The overall arm sizes and relative proportions are set to facilitate ease of handling and steady sighting, so that a stable scale reading can be taken.

Thus, more specifically:

Figure 3A shows a dual cross-arm (measuring) device, disposed with a support arm initially stood generally upright, or vertical, upon a support surface (i.e. the ground) - and a suspended pendulum, or weighted plumb line, intersecting the arcuate scale, calibrated to give a reading of some 90 degrees.

Figure 3B shows the support arm (progressively) pivoted about a lower end fulcrum, so that the sighting arm - set transversely (in fact orthogonally) thereto - aligns with a target [object].

Sighting - from initial approximation to a final determination - is facilitated and steady or stabilised by the co-operatively disposed support arm.

The pendulum, or plumb line, intersects the arcuate scale, allowing a reading to be taken, of the inclination of the sighting arm - and which, if correctly aligned, is equivalent to target angular elevation.

Figures 4A and 4B show a dual, differential-length cross-arm (measuring) device, with a freely-rotatable, bias-weighted arcuate scale, pivotally mounted at the cross-arm intersection.

An index (e.g. mark, pointer or graticule) on the support arm, allows scale readings to be taken, for a given relative disposition of the sighting arm.

Thus, more specifically:

Figure 4A shows a dual cross-arm (measuring) device, initially disposed with a support arm stood generally upright or vertical, upon a support surface (i.e. the ground) - and so with a support arm index, indicating some 90 degrees on the rotatable scale.

Figure 4B shows the dual cross-arm device, pivoted about a fulcrum, at the lower end of the support arm, so that the sighting arm axis aligns with a target.

The support arm index, indicates, on the scale, the angle of inclination of the sighting arm.

If correctly aligned, this is equivalent to the target angular elevation.

Figures 5A and 5B show a (measuring) device, of dual cross-arm configuration, as in Figures 3A and 3B, with a fixed arcuate scale and freely suspended pendulum, or weighted plumb line, and in which the upper end of the longer support/sight arm is used to sight the target object.

Thus, more specifically:

Figure 5A shows a rigid dual cross-arm (measuring) device, disposed initially with the (longer) support/sighting arm generally upright or vertical, and with a pendulum, or weighted plumb line, intersecting the arcuate scale at some 90 degrees.

Figure 5B shows the dual cross-arm device, pivoted about a fulcrum, at the lower end of the support arm, so that the upper, pointer, end of the (role-reversed) support or sighting arm aligns with a target object.

In this instance, the (transverse) former sighting arm, acts as a scale support.

The pendulum, or weighted plumb line, intersects the arcuate scale, allowing a reading to be taken, of support arm inclination.

If correctly aligned, this allows determination of target angular elevation - through the overall device and sightline geometry.

Figures 6A and 6B show an alternative (survey) use for a dual cross-arm (measuring) device - such as in Figure 3A and 5A - for determining angle of inclination, slope, gradient, fall or pitch, of either remote or adjacent target object, such as a building structure or (hard) landscape feature.

Thus, more specifically:

Figure 6A shows a rigid dual cross-arm (measuring) device, pivoted about a fulcrum at the lower end of the 'support' arm, so as to align the 'support' arm with the slope of a remote (albeit 'near distance') target object.

That is, the support arm can also be used for sighting.

Figure 6B shows a rigid, cross-arm (measuring) device, pivoted about a fulcrum, at 'support' arm lower end, so that the (transverse) 'sighting' arm is aligned, parallel with the slope of the remote object.

The pendulum, or weighted plumb line, intersects the fixed (arcuate) scale, at an angle, which if correctly aligned, allows target angular inclination, pitch or slope to be determined.

Figures 7A through 7C show alternative variant configurations for the sighting arm.

Thus, more specifically:

Figure 7A shows a sighting arm, with cross-sectional profile incorporating an (upper) surface 'V'-section groove or trough as the sight-line.

Figure 7B shows a sighting arm, surmounted by cross-hair aiming sights at each end of the sight-line.

Figure 7C shows a hollow sighting arm, with cross-hair sights at each end.

Figures 8A through 8C show alternative scale configurations for use with the (measuring) device described previously.

Thus, more specifically:

Figure 8A shows a linear scale, with parallel, linear graduations.

Figure 8B shows a linear scale, with (radially) angled graduations.

Figure 8C shows an arcuate quadrant scale.

Figure 9A and 9B depict variants of the (measuring) device of Figure 3A - the larger the scale, generally the more accurate the angle determination.

Thus, more specifically:

Figure 9A shows a hand-held version of the cross-arm (measuring) device, as in Figure 3A. Figure 9B shows a larger version of the dual cross-arm device, of Figure 3A, stabilised and braced by a support tripod - with a locking swivel mount.

The device is configured to be the size of a person, accommodating a scale of some 90 cm in length.

This in turn allows accuracy to some one minute of a degree - being equivalent, in navigational terms, to 6 nautical miles of longitude.

Figure 10A and 10B show how a dual cross-arm (measuring) device can be deployed in navigation - to determine latitude, by determining position in relation to the North Pole star.

Thus, more specifically;

Figure 10A shows how to locate the north pole star, currently Polaris, in the sky, by first locating the so-called Plough or Big Dipper constellation.
Figure 10B shows a dual cross-arm (measuring) device sighted on the North Pole star, allowing latitude to be determined.

Figures 11A and 11B show a dual cross-arm (measuring) device deployed in navigation - to calculate latitude or longitude, by determining position in relation to the sun.

Thus, more specifically;

Figure 11A shows a 'viewer' orientating a dual cross-arm (measuring) device, such that no 'lateral' shadow arises, and the device sighting line projects the light of the sun to the centre of the cast shadow.

If correctly aligned, the angle of inclination of the sighting arm, allows the elevation of the sun to be determined.

Figure 11B shows the dual cross-arm (measuring) device orientated so that a 'V' sight on the sighting arm is in line with its shadow - this being the correct alignment from which to determine sun elevation.

Figures 12A through 12C show a dual cross-arm (measuring) device deployed in navigation - to calculate longitude, by determining position in relation to one or more celestial bodies.

Thus, more specifically;

Figure 12A shows the dual cross-arm (measuring) device, orientated such that one member of the sight arm is sighted on due north (on the north pole star), the opposing end of the sight arm indicates due south.

Figure 12B shows the dual cross-arm (measuring) device, orientated at right angles to north and south - with the cross-arm pointing east to west, sighted on a target celestial body.

Figure 12C shows the 'viewers' position, as determined by the reading in Figure 12A, in conjunction with the time and an almanac, transposed onto a map.

A 'notional' inscribed circle (about the plumb line), of radius is depicted to represent error factor or accuracy, associated with scale reading resolution - with one degree equivalent to some 60 nautical miles, or one minute to some 6 nautical miles.
A second circle represents a subsequent reading 'spread', taken on the same celestial body.

The 'notional' circle overlap gives a good indication of a 'viewer/observer's' actual or true position.

Figure 13 shows a (measuring) device deployed as an aid to pilotage - to determine distance of a vessel from a target object, such as a lighthouse or charted mountain.

Figures 14A and 14B show the (measuring) device deployed to determine the pitch of a roof or canopy structure - as a building-surveying example.

Thus, more specifically:

Figure 14A shows a longer support arm, of a dual cross-arm (measuring) device, disposed parallel with the roof pitch of a remote (albeit 'near-distance') building.

Figure 14B shows a shorter (transverse) sighting arm, of a dual cross-arm (measuring) device, disposed parallel with the roof pitch of a remote building.

A pendulum, or weighted plumb line, intersects the fixed (arcuate) scale, at an angle, which (if correctly aligned), allows target angular inclination or slope to be determined.

Figure 15 shows the (measuring) device deployed to set levels, say on a construction site, or (hard) landscape contour survey.

Figure 16A through 16C show a collapsible, differential-length, dual cross-arm (measuring) device - foldable into a somewhat more compact linear structure, for carriage and storage.

Thus, more specifically:

Figure 16A shows an erect, functional rigid dual cross-arm (measuring) device, with fixed (linear) scale and freely suspended pendulum, or weighted plumb line.

Figure 16B shows the (linear) scale, pivoted about its point of contact with the support arm, to lie over the support arm.

Figure 16C shows the (transverse) sighting arm pivoted about its intersection with supporting arm, to lie behind the supporting to form a linear structure.

Figure 17A through 17E shows a collapsible cross-arm (measuring) device, as in Figures 16A through 16C, with telescopic arms.

Thus, more specifically:

Figure 17A shows an erect, rigid dual cross-arm (measuring) device, with telescopic arms, a fixed (linear) scale and freely suspended pendulum or weighted plumb line.

Figure 17B shows the scale pivoted, about its point of contact with the support arm, to lie over the support arm.

Figure 17C shows retraction of the telescopic support arm.

Figure 17D shows pivoting of the (transverse)-sighting sighting arm, about its intersection with the support arm, to lie behind the support arm.

Figure 17E shows retraction of the telescopic sighting arm, to produce a compact linear structure.

Figure 18A through 18C shows a (measuring) device with continuous circular (360 degree) scale.

Thus, more specifically:

Figure 18A shows a (measuring) device comprising a unitary sighting and support member, with a pivoted, freely-rotatable, bias-weighted circular scale, and an index on the member for scale reading.

Figure 18B shows the unitary member aligned with the target object.
The member index indicates angle of inclination of the sighting member - and which, if correctly aligned, allows target angular inclination to be determined.

Figure 18C shows a dual cross-arm (measuring) device with weighted circular scale, and an index on a support arm, to indicate device inclination.

Sighting cross-hairs, grids or graticules are located on the sighting arm.

Figures 19A and 19B show a circular (measuring) device, with integrated internal support and sighting.

Thus, a circular weighted scale is enclosed in a shallow-depth, cylindrical diskhousing.

A viewing window, with index mark, allows scale reading.

Diametrically-opposed graticules are pivotally mounted in the housing and can be deployed as upstands from the surface, for target sighting, or folded way for transport and storage.

In an alternative embodiment (not shown) the casing and internal wheel or drum may of sufficient depth to allow the sights to be incorporated within the rim - so that a 'viewer' would effectively look 'through' the device, in order to sight a target.

A trigger release allows movement of the scale, upon sighting of the target, and positive damping or braking of the scale, in order to steady and lock the reading - whereupon the same reading can be viewed with the device no longer sighted on the target.

The scale would be released when ready for the next sighting.

The same, or a separate trigger release, could be used to deploy or retract the sighting upstands. Thus, more specifically;

Figure 19A shows the circular (measuring device) in a compact configuration, before use.

Figure 19B shows the circular (measuring) device, with sighting graticules hinged at 90 degrees to the housing, and sighted on a target.

Referring to the drawings, a (measuring) device 10, is adapted for use as a tool for navigation or surveying, to determine elevation or (in)clination.

The device 10 comprises a rigid, elongate member 11, the distal end employed as a pointer 12, to sight a target object 30, and the proximal end 14, serving as fulcrum about which member 11 can pivot.
Operationally, rather than haphazard ad hoc sighting, some procedural framework for sighting is helpful for consistent readings.

Thus, for example, the device is conveniently initially stood generally upright, or vertical, and progressively leaned or inclined towards the target.

A scale 22, 28 is mounted upon the member 11 intermediate the distal pointer and fulcrum ends.

In order to sight the target 30, a 'viewer' (not depicted) stands alongside the device 10, facing the scale 22, 28, and pivots the device 10 laterally, until the pointer end 12, of the rigid member 11, aligns with the target 30.

In order to determine the degree of inclination of the device 10, and hence the inclination or elevation of a target 30, the device must incorporate a scale.

The scale can be variously configured, including:

- an (arcuate or linear) scale 22, secured to member 11, with a pendulum or weighted plumb line 24, freely suspended from member 11 above the scale 22, for reading at the scale intersection; or

- a weighted arcuate scale 28, pivotally connected 26, to member 11, with an index 29, for reading on scale 28.

Both the pendulum/weighted plumb line 24 and weighted scale 28 continuously self-align, under the influence of gravity, to point towards the centre of the Earth.

Thus, the degree of inclination of the vertical member 11, is determined by noting the intersection of the scale 22, 28, by the pendulum/plumb 24, or index 29, respectively.

Providing the target object 30 is correctly sighted, its inclination or elevation can be determined from the degree of inclination of the unitary member 11.

In an alternative configuration, depicted in Figures 3A through 6B, a (measuring) device is configured as a (dual) cross-arm structure 15 - specifically, a [upright] support arm 19, and a [transverse] sighting arm 17.

The cross-arm structure 15 also incorporates a scale, configured as either:

- a 'fixed' scale 22, interposed between cross-arms 17, 19 - with a free-swinging pendulum, or weighted plumb line 24, pivotally attached at the cross-arm intersection 25, for reading at the scale intersection.

- a bias-weighted and freely-rotatable scale 28, pivotally attached to the cross structure 15, at the cross am 17, 19 intersection 25 - with an index 29 on the cross member 19, for scale reading.

The cross-arm structure 15, can be employed in various ways to determine elevation or inclination.

For example, a sighting arm 17, may act as the 'line of sight' 34, for viewing a target object 30.

A 'viewer' (depicted in Figures 9A through 9B) would stand in line with the sighting arm 17, at a view point 32, and pivot the support arm 19, backwards and forwards, about its fulcrum 14, until the sighting arm 17 (longitudinal axis) and target object 30 are aligned.

Sighting errors could arise by relying solely upon one distal end of the sighting arm.

Rather, its longitudinal span must be viewed - either from one end to another, or by standing alongside and extrapolating the longitudinal axis to the target.

Target 30 sighting can be facilitated by incorporating a continuous groove 42 in the upper surface of the sighting arm 17, or by mounting sighting cross-hairs or graticules 43, upstanding at opposite ends thereof.

Alternatively, a hollow sighting arm 45 allows sighting by viewing internally along the (entire) tube length, from one end to another.

Inclination of the sighting arm 17, once aligned with a target 30, is determined by taking a reading from the scale 22, 28 - so that, if correctly sighted, target 30 angular elevation can be determined.

If more convenient for a given viewing situation and target, the respective roles of otherwise 'dedicated' sighting and support arms 17, 19 could be reversed or interchanged.

That is, the distal end of the support arm 19 could itself be sighted upon the target 30.
A 'viewer', positioned alongside, would then pivot the entire device 15, about the fulcrum 14, until the distal, pointer, end 12 of the support arm 19 aligns with the target 30.

A scale reading - representing inclination of the support arm 19, is taken where the pendulum/plumb 24 intersects the scale 22.

With a correct sighting, target 30 inclination or elevation can be determined.

The device 15 is also applicable to surveying - say, to sight a target 30 such as a [distant] roof pitch, gradient or slope, with one of the cross-arms 17, 19, set parallel with the target slope line.
To achieve this, a 'viewer' could stand alongside the device 15, facing the scale 22, 28, and pivot the device about fulcrum 14, until either arm 17, 19, is sighted parallel to the target object.

A reading can then be taken from the scale 22, 28.

Given correct sighting alignment, target 30 inclination can be determined.

Another practical application of the measuring device 10 or 15, is as a simple and reliable navigational aid.

More specifically, location can be derived by using the device to determine elevation of celestial bodies.

For example, latitude can be determined by calculating the angle between a position on the Earth's surface and the north pole star.

In order to locate the north pole star, currently Polaris, the constellation known as The Plough or The Bigger Dipper must first be located.

The Plough constantly revolves around Polaris in an anti-clockwise direction, when viewed in northern latitudes.

At the outer edge of The Plough are two stars known as 'pointer stars'.

By following the line of the pointer stars, the next star seen is Polaris - the north pole star (depicted in Figure 10A).

By sighting one arm at the pole star, and reading 'indicated degrees' from the scale, latitude can be determined directly.

For an observer taking a sighting, and making an attendant angular elevation determination, from the standpoint of the Earth' surface or circumference - rather than a geometric centre of the Earth - the great distance of the star - even in relation to the Earth's radius - effectively makes any parallax viewing error(s) insignificant.

The cross instrument configuration, and opposed scale mounting, allow measurement of the opposite angle to the actual angle between the star and Earth.

The smaller resulting angle is equivalent to some 90 degrees of latitude from equator to pole.

Having determined true north, by sighting the pole star (in the northern hemisphere), true south may also be determined, in that the opposed end of the sight arm will point exactly true south.

In addition to sighting the pole star, navigation can be aided by sighting other celestial bodies, for example to determine longitude.

When viewed from the standpoint of an observer at the Earth's surface, celestial bodies, travel or pass from east to west (until lost or eclipsed by the viewing horizon) - on a so-called 'ecliptic' path across the horizon, in an orderly and predictable manner - thereby achieving given perceived 'ascension' and 'declination' angles, at particular locations and times.

Accurate observation of one, or several of these celestial bodies, in conjunction with an appropriate almanac and accurate time-piece, allows longitude to be determined.

As the Earth orbits around the sun, observed stars appear to move westward every night, at the same observation time, by some 0.98 degrees.

That is, their position, over a given point on the Earths surface moves west 58.8 nautical miles.

Simultaneously, as the Earth progresses in its orbit around the sun, and seasons change due to the Earth's 'angle of obliquity' to the ecliptic (path or plane), the elevation or azimuth of the stars also changes, in ascension (apparently rising) or declination (apparently falling or descending).

The overall characteristic pattern of progress is such that, midwinter (the Solstice) is the maximum ascension of the star due south in the sky, and midsummer (the Solstice) is the minimum declination.

For example - if a given star is predicted, by an almanac, to be at an angle of 23.8 degrees above the equator, exactly due south at 0 (zero) degrees longitude at 24.00 hours GMT - it is a relatively straightforward exercise to find an observer's position on the Earth's surface, to within some 6 to 12 nautical miles, using a (measuring) device according to the invention.

By employing the preferred configuration of dual, differential-length, cross-arm (measuring) device, of certain embodiments of the invention, an observer's viewing location can be determined through the following sequence of steps:

- latitude is determined by sighting the cross/sighting arm on the (north) pole star;

- the 'viewer-observer' finds true south (opposite direction to the pole star);

- the 'viewer-observer' identifies the star, and its 'projected' ascension at midnight, and deducts latitude 'made good' (ie through a correction factor);

- the 'viewer-observer' then rotates the (measuring) device by some 90 degrees, so that it is at right angles to a north and south axis, and with the sighting cross-arm pointing east to west.

- the 'viewer-observer' now measures the difference in degrees between the observed position of the star, against its predicted position at 0 (zero) degrees longitude.

- by correcting the difference between the predicted, and the actual, angle of the star, into degrees and minutes, east or west of 0 (zero) degrees reference datum line (or meridian), the 'viewer-observers' longitude can be determined.The same rules apply to lunar observation, although the moon is only observable at certain periods of the month.

If a large (measuring) device is employed, with say a 90 centimetre rule,1 millimetre is equivalent to 1 minute of arc - which in longitude represents an accuracy to some 6 nautical miles. However, with a small hand held device, the accuracy will be to one degree, equal to 60 nautical miles.

Error can be reduced by taking further sightings, over a period of time, and transposing the results upon a chart, as circles measuring 6 or 60 miles, as appropriate.

The (segmental) intersection or overlap of such circles gives a better - and overall good - indication of position.

The (measuring) device can also be deployed to take solar sightings.

The elevation of the sun 52, in conjunction with appropriate almanacs and an accurate timepiece, allows both latitude and longitude to be determined.

Since it is not possible to view the sun directly with the device, an indirect measurement is taken.
For example, a dual, differential-length, cross-arm (measuring) device 15 is orientated such that its shadow 37 is cast on the ground or a horizontal surface.

When no lateral shadows of the device arms 17, 19 are visible, and the V sight 42, or sighting tube 45, projects the light of the sun 52 to the centre of the shadow 37, the device is correctly orientated and the inclination of the sight arm 17 can be read from the scale, allowing the elevation of the sun 52 to be determined.

The (measuring) device 10, 15 may also be used in pilotage, to determine the distance of a vessel 51 from a known target object 30, such as a lighthouse or a charted mountain - the height of which is known.

By sighting an arm of a (measuring) device 10, 15 , on the target object 30, its elevation can be determined.

Using basic (triangulation) geometry, the elevation and height of an object can be used to calculate the distance to the object.

To avoid error, account must be taken of height above sea level and distance to the horizon.

Whether configured as a unitary member 10, or a dual cross-arm structure 15, the (measuring) device, must be robust and durable.

As such, construction from materials such as, wood, synthetic plastics (moulded or extruded), or metal are envisaged.

Component List

10	(measuring) device
11	unitary sighting member
12	(distal) pointer end
14	fulcrum
15	(dual) cross-arm (measuring) device,
17	sighting arm
19	support arm
22	(fixed) scale
24	pendulum/weighted plumb line
25	cross-arm intersection
26	pivot
28	weighted scale
29	index
30	target (object)
32	viewpoint
34	line of sight
35	circular (measuring) device
37	shadow
38	shadow of sighting groove

42	groove/channel
43	sighting graticule
45	hollow (sighting) arm
48	tripod
49	trigger release
51	vessel
52	sun

Claims

{unitary member + pendulum/plumb}

A (measurement) device (10)
comprising
an elongate unitary member (11),
[one end] acting as a pointer (12),
for sighting upon a target (30),
by viewing, either along its length,
from the opposite end,
or from alongside,
with the other end as a support fulcrum (14);
a fixed scale (22), mounted upon the member,
a pendulum or weighted plumb line (24),
freely suspended from the member,
to intersect the scale,
allowing (direct) scale reading,
for determining member inclination,
and thus target elevation.

{cross-arm - pendulum/plumb}

A (measurement) device (15),
as claimed in Claim 1
with a plurality of intersecting arms (17, 19),
disposed in a cross configuration,
one arm (17) deployable as sight line,
for line-of-sight viewing longitudinally,

from one end to another,
in alignment with a remote target object (30);
another arm deployable as a fulcrum (14),
to provide stability, and
to act as pivot,
about which to adjust sight-line orientation,
inclination and/or elevation;
a pendulum or weighted plumb line (24),
freely suspended from the arm intersection (26);
and a scale (22),
interposed between the arms (17, 19),
below the suspended pendulum, or plumb line (24),
to allow a direct scale reading,
for determining sight-line inclination, or elevation,
of a target (30).

{Differential arm length}

A (measurement) device,
as claimed in either of the preceding claims,
with differential length arms,
respectively or interchangeably,
co-operatively deployable for sighting and support.

{weighted scale}

A (measurement) device,
as claimed in any of the preceding claims,
with a plumb-weighted scale (28),
pivotally attached to either a unitary member (11),
or a cross-arm intersection (25);
an index mark upon one member or arm,
alignable with a scale subdivision,
for determining target inclination, or elevation.

{arcuate scale}

A (measurement) device,
as claimed in any of the preceding claims,

with an arcuate scale,
subdivided for direct reading of
target angular elevation, or inclination.

{circular scale}

A (measurement) device,
as claimed in any of the preceding claims,
with a generally circular scale,
(subtending some 360 degree span),
subdivided for direct reading of
angular elevation, or inclination.

{linear scale}

A (measurement) device,
as claimed in any of the Claims 1 through 4,
with a linear scale,
subdivided for direct reading of
target angular elevation, or inclination.

{adjustable arm length}

A measurement device,
as claimed in either of the preceding claims,
with an adjustable length arm or member.
{sight line - groove}

A (measurement) device,
as claimed in any of the preceding claims,
incorporating a continuous groove,
along a longitudinal side edge of an arm or member,
to facilitate taking a sight line to a target.

{sight line - sights}

A measurement device,
as claimed in any of the preceding claims,

with upstanding graticules upon an arm or member,
to facilitate taking a sight line.

{sight line - hollow}
A measurement device,
as claimed in any of the preceding claims,
with a hollow arm or member,
to facilitate taking a sight line.

{tripod}

A (measurement) device,
as claimed in any of the preceding claims,
with a mounting stand, such as a tripod.

{illustrated embodiments}

A (measurement) device,
substantially as hereinbefore described,
with reference to, and as shown in,
the accompanying drawings.

{method}

A method of determining
target inclination/declination, or elevation,
comprising the steps of target alignment,
by pointing either one end, or the entire length of,
one of a pair of intersecting arms,
co-operatively and interchangeably disposed,
as either pointer or support,
in a cross configuration,
with one end in ground contact, as a fulcrum,
about which the pointer arm is pivoted,
for target sighting,
suspending a pendulum from the intersection
of the pointer and support arms,

mounting a scale between those arms,

and taking a direct reading

of target angular inclination/declination,

from the intersection of the pendulum with the scale.

A device (15), for angular determination or measurement, employs a cross-arm structure (17, 19), with a sighting arm for alignment with a target (30), either viewed along it, or from one side; the angle of inclination of the arm being determined by either a pendulum or plumb line (24), freely suspended from the arm intersection (26); a scale (22) interposed between the arms (17, 19); a scale reading being taken where the pendulum or plumb line intersects the scale; or, alternatively, a plumb weighted scale (28), is pivotally mounted at the intersection of the cross-arms (26), with an index upon one cross-arm (17, 19) for scale reading, whereby, if correctly sighted, the scale reading reflects target angle of elevation or inclination .

(Figure 3B)

312

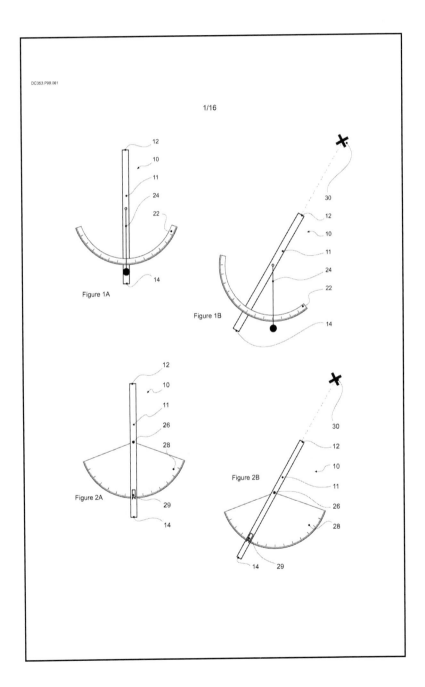

Figure 1A

Figure 1B

Figure 2A

Figure 2B

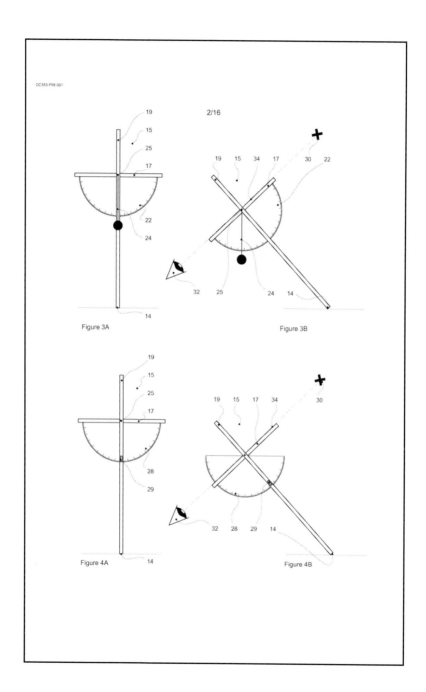

Figure 3A

Figure 3B

Figure 4A

Figure 4B

314

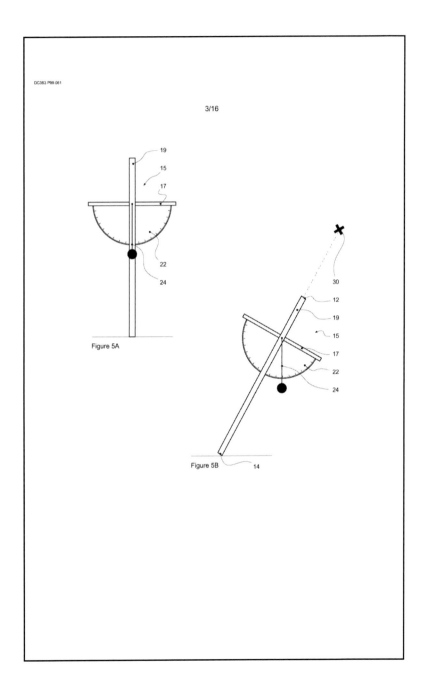

Figure 5A

Figure 5B

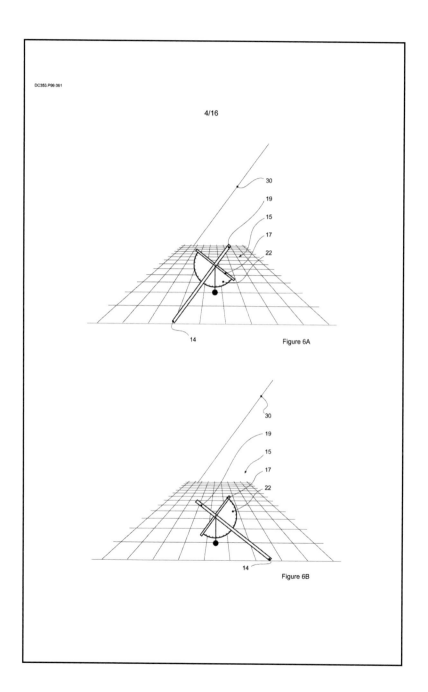

Figure 6A

Figure 6B

316

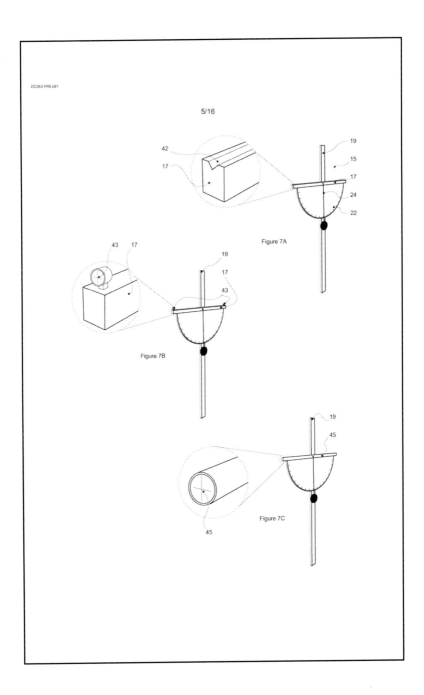

Figure 7A

Figure 7B

Figure 7C

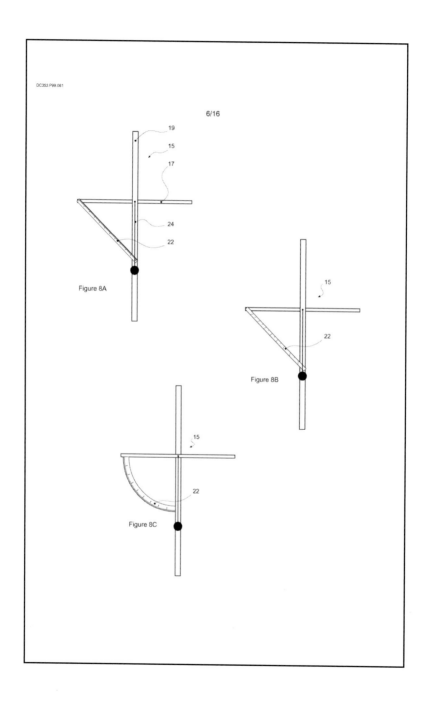

6/16

Figure 8A

Figure 8B

Figure 8C

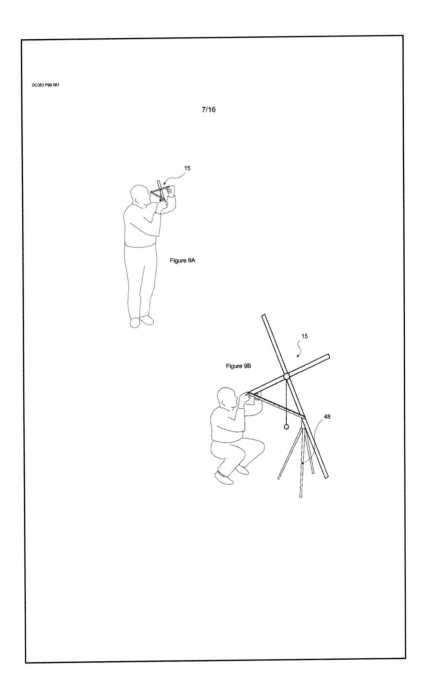

Figure 9A

Figure 9B

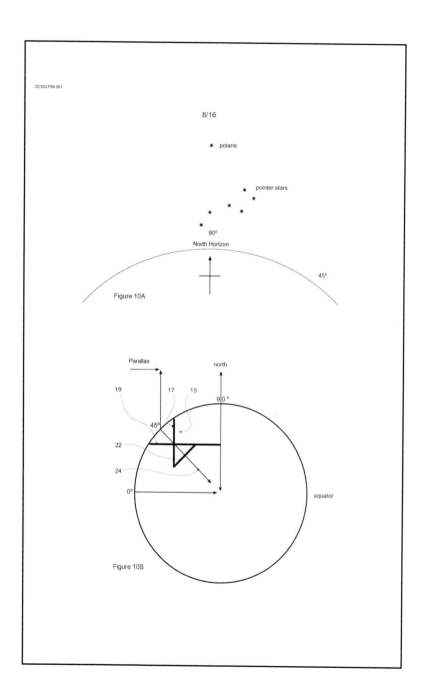

8/16

* polaris

* pointer stars

90°

North Horizon

45°

Figure 10A

Parallax

north

19   17   15

45°   90°

22

24

0°

equator

Figure 10B

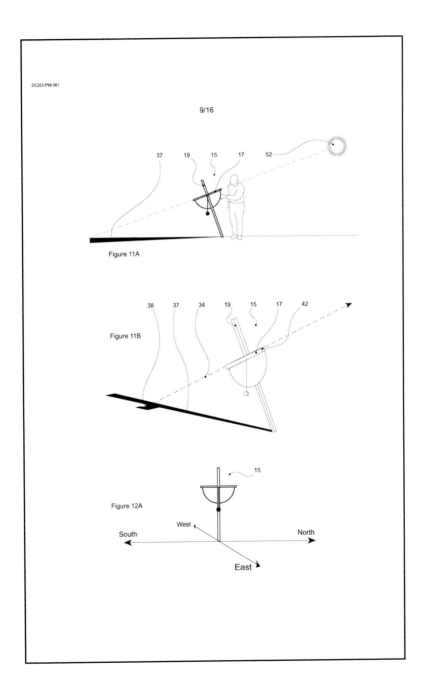

Figure 11A

Figure 11B

Figure 12A

South

West

North

East

DC353.P99.061

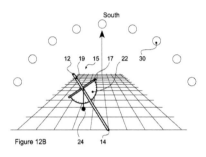

Figure 12B

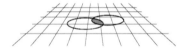

Figure 12C

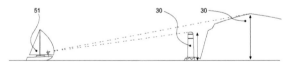

Figure 13

322

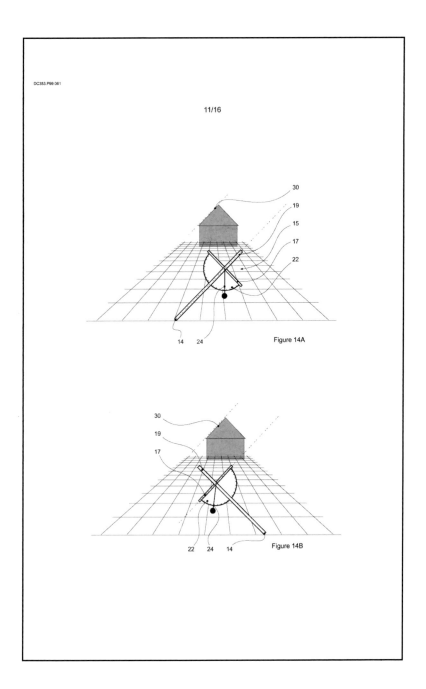

Figure 14A

Figure 14B

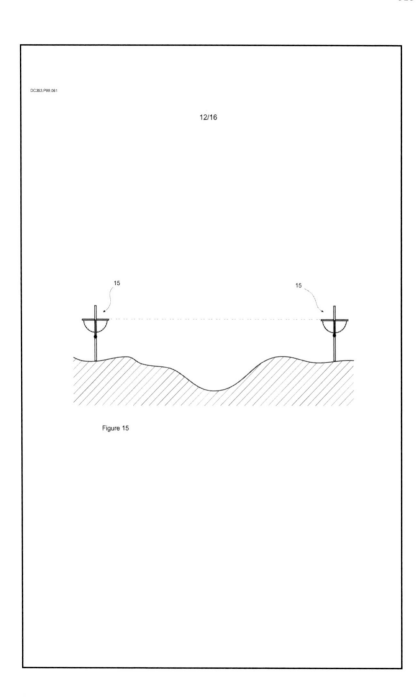

Figure 15

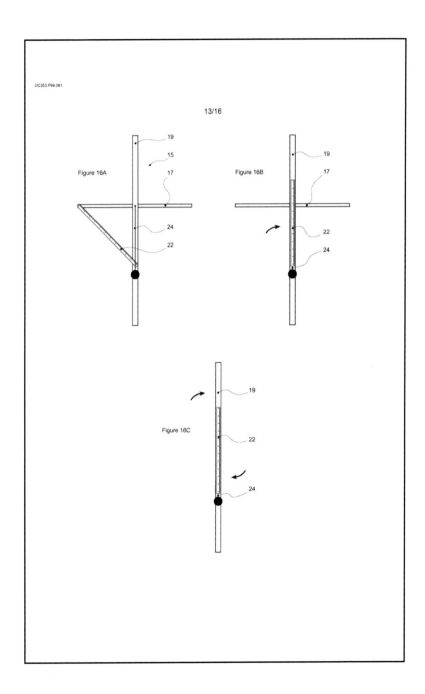

13/16

Figure 16A

Figure 16B

Figure 16C

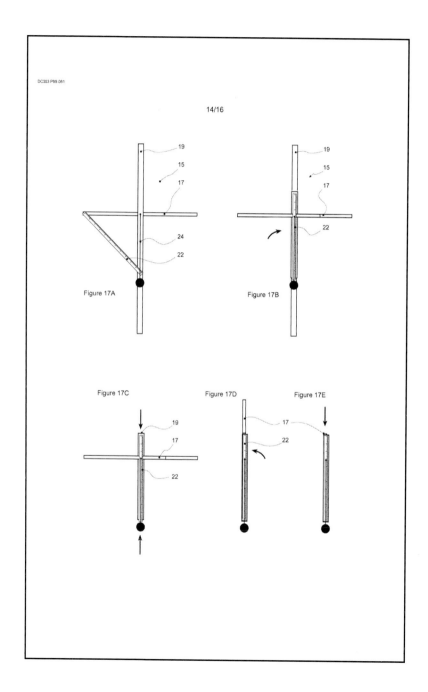

Figure 17A

Figure 17B

Figure 17C

Figure 17D

Figure 17E

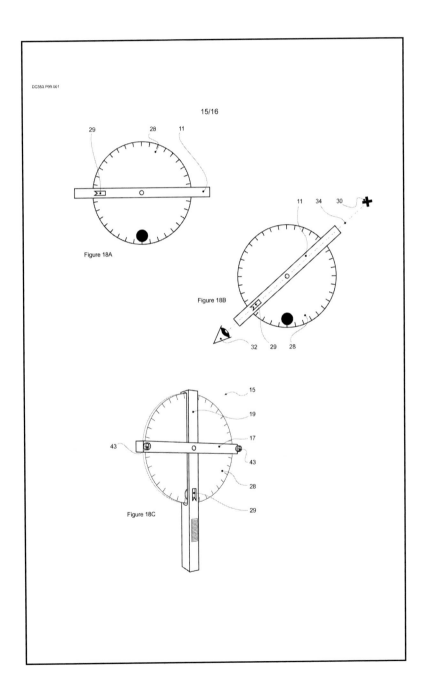

Figure 18A

Figure 18B

Figure 18C

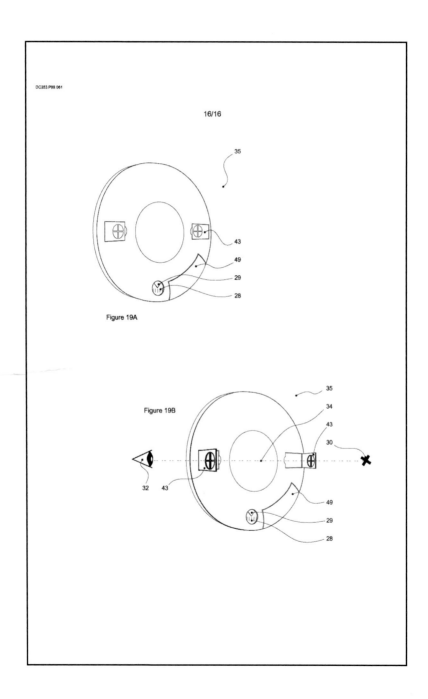

16/16

35

43

49

29

28

Figure 19A

35

34

43

30

32   43

49

29

28

Figure 19B

REFERENCES

i   www.crichtonmiller.com

ii   J. Bowles. The Gods, Gemini and the Great Pyramid. ISBN 0-9666371-1-9.

iii   Charles. H Hapgood. Earth's Shifting Crust, Pantheon Books, 1958.

iv   Erika Cheetham. The Further Prophesies of Nostradamus.
Corgi Books, ISBN 0-552-12299-8.

v   The Good News Bible. The book of Matthew. Page 4.2.
ISBN No 0-564 00311-5

vi   Swami Prabhupada. Bhagavad Gita. Page 29. Plate 6. ISBN 0-912776-80-3.

vii   Robert Brydon. The Masons and The Rosy Cross. ISBN 0-9521493-3-8.
Arms of Masons.

viii   Tablet 1 of the Emerald Tablets of Thoth.
http://members.tripod.com/dragoncaer/id75.htm

ix   The Good News Bible. The book of Matthew. 1.2.13 ISBN No 0-564 00311-5

x   Robert Brydon. The Masons and The Rosy Cross. ISBN 0-9521493-3-8.
What is a Freemason?

xi   Swami Prabhupada. Bhagavad Gita. Page 144. Plate 12. ISBN 0-912776-80-3.

xii   Lorraine Evans. The Kingdom of the Ark. ISBN 0-684-86064-3. Pages 28-29.

xiii   The Good News Bible. Genesis 7.10. ISBN No 0-564 00311-5

xiv   Miguel Angel Vergara Calleros. http://www.1spirit.com/kukuulkaan/page6.html
Chichen Itza Astronomical Light and Shadow Phenomena of the Great
Pyramid.

xv   Maureen Palmer. http://www.godsintherocks.com

xvi   The Earl of Rosslyn. A Short History of Rosslyn Chapel. Rosslyn Chapel Trust,
Page 27.

xvii   Maurice Cotterell. The Tutankhamun Prophesies. ISBN 0-7472-2216-9. Page
23.

xviii   Maurice Cotterell. The Tutankhamun Prophesies. ISBN 0-7472-2216-9. Page
23.

xix   Andrew Collins. Gateway to Atlantis. ISBN 0-7472-2280-0.

xx   Robert Bauval & Adrian Gilbert, The Orion Mystery. ISBN 0-7493-1744-2

xxi   Royal Yachting Association. Navigation anD RYA Manual. ISBN 0-7153-8630-
1. Page 13.

xxii   Maurice Cotterell. The Tutankhamun Prophesies. ISBN 0-7472-2216-9. Page
58.plate 45a.

xxiii   Robert Bauval & Adrian Gilbert, The Orion Mystery. ISBN 0-7493-1744-
2.Pages 238-250.

xxiv   Robert Bauval & Adrian Gilbert, The Orion Mystery. ISBN 0-7493-1744-2.

xxv   Rudolf Gantenbrink. World Wide Web as the Uphaut Project
http://www.cheops.org. The Findings.

xxvi   Practical Boatowner. www.pbo.co.uk. Issue No 412, Boat craft. Page 117.

xxvii   www.rostau.com

xxviii Robert Bauval & Adrian Gilbert, The Orion Mystery.
        ISBN 0-7493-1744-2.Page 238.

xxix   Lorraine Evans. Kingdom of the Ark. ISBN 0-684-86064-3. Page 14.

xxx   Lorraine Evans. Kingdom of the Ark. ISBN 0-684-86064-3. Page 152.

xxxi   Royal Yachting Association. Navigation an RYA Manual.
        ISBN 0-7153-8630-I. Page 63.

xxxii   Dava Sobel, Longitude. ISBN 1-85702-502-4.Page 120.

xxxiii Dava Sobel, Longitude. ISBN 1-85702-502-4 Page 60.

xxxiv   Robert Bauval & Adrian Gilbert, The Orion Mystery.
        ISBN 0-7493-1744-2 Page 215.

xxxv   Rudolf Gantenbrink. World Wide Web as the Uphaut Project
        http://www.cheops.org.

xxxvi   Sir William Flinders Petrie. The Pyramids and Temples of Gizeh 1883.
        Sections 136-139

xxxvii Sir William Flinders Petrie. The Pyramids and Temples of Gizeh 1883.
        Sections 129-135

xxxviii http://www.wfu.edu/~cyclone/

xxxix   Dr Covey's Home page http://www.wfu.edu/~cyclone/

xl   http://www.scotland-info.co.uk/c-nish.htm

xli   http://www.northcoastal.freeserve.co.uk/wooden_or_sea_henge.htm

xlii   http://www.rubens.anu.edu.au/student.projects/tajmahal/home.html

xliii   http://www.museum.mq.edu.au/eegypt2/djoser1.html

xliv   http://guardians.netegypt/saqqara1.html

xlv   http://encarta.msn.com/find/Consise.asp?ti=00676000

xlvi   Robert Bauval & Adrian Gilbert, The Orion Mystery.
        ISBN 0-7493-1744-2.Page 217 plate 19.

xlvii   http://encarta.msn.com/index/consiseindex/4E/04E5A000.htm

xlviii Robert Bauval & Adrian Gilbert, The Orion Mystery.
        ISBN 0-7493-1744-2.Page 201.Plate 17.

xlix   http://www.aquarian-age.net/

l   http://www.christopher-saint.com/house_of_pisces.htm

li   Clive Ruggles. Astronomy before the Telescope. ISBN 0-7141-1746-3,
        Archaeoastronomy in Europe. pages 17-27

lii   http://www.humanities.ccny.cuny.edu/history/reader/weighingheart.htm

liii   Ronald A. Wells. Astronomy before the Telescope. ISBN 0-7141-1746-3,
        Astronomy in Egypt. pages 28-41

liv   Ronald A. Wells. Astronomy before the Telescope. ISBN 0-7141-1746-3,
        Astronomy in Egypt. Pages 28-41. plate 12

lv   Discovering Archaeology. Issue 2.March/April 2000. Mediaeval Craftsmen
        Rivalled Descartes.

lvi   R Eisenman & M.Wise. The Dead Sea Scrolls Uncovered.(4Q321)

lvii   R Eisenman & M.Wise. The Dead Sea Scrolls Uncovered.(4Q318) (plate 23)

lviii   Philip's Planisphere. For latitude 51.5° North.

lix   Erling Haagensen & Henry Lincoln. The Templars Secret Island. Page 82.
         ISBN 1-900624-37-0.

lx   Erling Haagensen & Henry Lincoln. The Templars Secret Island. Page 83.
         ISBN 1-900624-37-0.

lxi   Rudolf Gantenbrink. World Wide Web as the Uphaut Project
         http://www.cheops.org.

lxii   Sir William Flanders Petrie. The Pyramids and Temples of Gizeh 1883.
         Surveying Methods.

lxiii   Manfred Lurker. Gods and Goddesses, Devils and Demons. 1995, Routledge.
         ISBN 0-415-03944-4

lxiv   A RYA Manual. Navigation. ISBN 0-7153-8631-X.Chartwork. Page 14.

lxv   A RYA Manual. Navigation. ISBN 0-7153-8631-X.Chartwork. Page 13.

lxvi   A RYA Manual. Navigation. ISBN 0-7153-8631-X.Chartwork. Page 14.

lxvii   David Rohl. Legend. The Genesis of Civilization.
         ISBN 0-7126-7747-X.Chapter 8.Pages252-253.

lxviii   Andrew Collins. Gateway to Atlantis. ISBN 0-7472-2280-0. Page 148. Plate 19.

lxix   Temple of Haphuset friezes. Egypt

lxx   Peter Whitfield. The Charting of the Oceans. ISBN 0-7123-0493-2.
         Page 6, the tomb of Hatshepsut stone reliefs.

lxxi   http: astrolabes.org/astrolabe.htm.

lxxii   http: astrolabes.org/astrolabe.htm.

lxxiii   http: astrolabes.org/astrolabe.htm.

lxxiv   Andrew Collins. Gateway to Atlantis. ISBN 0-7472-2280-0. Page 276. Plate 25.

lxxv   Peter Whitfield. The Charting of the Oceans. ISBN 0-7123-0493-2. Page 30.
         Biblioteca Apostolica in the Vatican.

lxxvi   Robert Dudley's illustration of the use of the Quadrant. The British Library.
         Maps C.8.d.9 (1).

lxxvii   Pisan Chart, Bibliotheque Nationale, Paris, France.

lxxviii A RYA Manual. Navigation. ISBN 0-7153-8631-X. Chartwork.Page 14.

lxxix   http:/www.kheraha.com The website of David Ritchie

lxxx   Christopher Knight & Robert Lomas. The Hiram Key.
         ISBN 0-09-969941-9.Pages 6-23.

lxxxi   http:/www.geocities.com/~jrancel/canlyend-eng.html .The Dark Ocean

lxxxii   A RYA Manual. Navigation. ISBN 0-7153-8631-X.Passage making. Page 148

lxxxiii House of Chacona, Museum, Tenerife.

lxxxiv   Andrew Collins. Gateway to Atlantis. ISBN 0-7472-2280-0. Page 148. Plate 6

lxxxv   Encarta World Atlas.

lxxxvi   http:/www.ing.iac.es/lapalma/history.html

lxxxvii Andrew Collins. Gateway to Atlantis. ISBN 0-7472-2280-0. Page 277. Plate 25.

lxxxviiiAndrew Collins. Gateway to Atlantis. ISBN 0-7472-2280-0. Page 46

lxxxix Microsoft Encarta. World Atlas 99.

xc   Andrew Collins. Gateway to Atlantis. ISBN 0-7472-2280-0. Pages 250-253

xci   Christopher Knight & Robert Lomas. The Hiram Key. ISBN 0-09-969941-9

xcii   Andrew Collins. Gateway to Atlantis. ISBN 0-7472-2280-0. Pages 250-253

xciii   Andrew Collins. Gateway to Atlantis. ISBN 0-7472-2280-0. Page 148 .Plate 19

xciv   Pisan Chart, Bibliotheque Nationale, Paris, France.

xcv   Christopher Knight & Robert Lomas. The Hiram Key. ISBN 0-09-969941-9
        Page 176 Plate 2.

xcvi   Erling Haagensen & Henry Lincoln. The Templars Secret Island.
        ISBN 1-900624-37-0.

xcvii   Erling Haagensen & Henry Lincoln. The Templars Secret Island. Page 10.
        ISBN 1-900624-37-0.

xcviii Peter Whitfield. The Charting of the Oceans. ISBN 0-7123-0493-2.

xcix   http://www.gci-net.com/users/v/vrartist/leadcrosses.html

c   http://www.gci-net.com/users/v/vrartist/leadcrosses.html

ci   Frank Waters. Book of The Hopi . Source Material by: Oswald White Bear
        Fredericks. ISBN 345-01717-X-125. Library of Congress Catalog No. 63-
        19606